unzipped genes

taking charge
of baby-making
in the new millennium

IN THE SERIES

America in Transition: Radical Perspectives

edited by Gary L. Francione

martine rothblatt

unzipped genes

taking charge of baby-making in the new millennium

temple

university

press

philadelphia

TEMPLE UNIVERSITY PRESS, PHILADELPHIA 19122

PUBLISHED 1997
PRINTED IN THE UNITED STATES OF AMERICA

∞ THE PAPER USED IN THIS PUBLICATION MEETS THE REQUIREMENTS OF AMERICAN NATIONAL STANDARD FOR INFORMATION SCIENCES—PERMANENCE OF PAPER FOR PRINTED LIBRARY MATERIALS, ANSI Z39.48-1984

TEXT DESIGN BY WILL BOEHM

LIBRARY OF CONGRESS CATALOGING-IN-PUBLICATION DATA

Rothblatt, Martine Aliana, 1954–

Unzipped genes : taking charge of baby-making in the new millennium / Martine Rothblatt.

p. cm. — (America in transition)

Includes index.

ISBN 1-56639-522-4 (alk. paper).—ISBN 1-56639-554-2 (pbk.: alk. paper)

1. Human reproductive technology—Moral and ethical aspects. 2. Human genetics—Law and legislation—United States. 3. Transgenic organisms. I. Title. II. Series.

RG133.5.R683 1997

176—DC20 96-35970

To my parents
Hal and Rosa Lee

and their parents
Goldie and Isadore, Jennie and Sam

contents

series editor's foreword

"Eugenics" is a word that often elicits the most negative images—and rightly so. After all, it was not long ago that the Nazis tried and failed to extinguish the genes of Jews, Gypsies, and others. And in 1927, the United States Supreme Court declared in *Buck v. Bell* that the eugenic sterilization of the "feebleminded" was constitutional. According to no less a jurisprudential luminary than Oliver Wendell Holmes, "The principle that sustains compulsory vaccination is broad enough to cover the Fallopian tubes. Three generations of imbeciles are enough." Indeed, eugenics is often associated with the image of Mary Shelley's Dr. Frankenstein, who, in seeking to use science to create "perfect" life, caused great tragedy for his creature as well as for society.

But this view of eugenics is both limited and limiting. The reality is that on some level, we *all* endorse *some* notion of eugenics. When parents adopt a child, they almost always specify certain characteristics, such as race, sex, and age. In choosing a mate, we implicitly make decisions about our offspring: When a Swedish person and a Chinese person choose to mate, they have made a decision about their preference for (or at least their tolerance of) a "mixed" child. Women have long used various means, ranging from consuming the penises of exotic animals to praying to using modern concoctions designed to increase alkalinity and decrease acidity in the reproductive tract, in order to increase the chances of having a male child. And sperm and egg banking, which allow the genetic consumer literally to choose from a menu of various demographic characteristics, are no longer all that controversial.

Personal eugenics has been with us a long time, and in general, the law does not intervene in personal eugenic choices. Blind people are at liberty to produce blind children. There are no laws that forbid procreation by obese or bald people, or by people who are likely to pass diseases on to their offspring. Although there may be some limits in some places on the use of artificial biotechnology (such as sperm and egg banks) to facilitate personal eugenic choices, these restrictions are often avoidable by those who want to use such technologies and have the money to seek alternatives. As long as parents do not cause "biological harm" to their offspring, we usually do not allow the law to interfere with personal eugenic choices. We may disagree about what constitutes "biological harm," but we generally agree that a basic notion of human rights *includes* the right to make at least certain decisions about the type of children that we have.

There is, however, new technology on the horizon that will make it possible for people—and governments—to have greater control than ever over these eugenic choices. In 1989, Congress initiated the Human Genome Project, which is being coordinated primarily by the National Institutes of Health and the Department of Energy. Although most funding for the project comes from the United States, governments around the world are involved in this worldwide research effort intended to map each human gene—of which there are an estimated 50,000 to 100,000—and to understand the function of each.

In most human cells, there is a genetic blueprint, or genotype, of the entire person. These cells each contain 23 pairs of chromosomes, which consist of deoxyribonucleic acid (DNA).

A molecule of DNA consists of two strands that resemble a ladder twisted into a spiral configuration that is referred to as a "double helix." Each strand of DNA consists of a linear arrangement of units called nucleotides, which are composed of one sugar, one phosphate, and a nitrogenous base. The two strands of DNA are held together by weak bonds between the nitrogenous bases, forming base pairs. There are some 3 billion base pairs of DNA. A gene is a specific sequence of these nucleotide bases, ranging from fewer than one thousand bases to several million. Each gene provides the information necessary to produce proteins that are responsible for individual human traits.

On one hand, the Human Genome Project has exciting potential: the ability to identify particular genetic abnormalities that occur when a deficient nucleotide composition that compromises a gene fails to produce the protein needed to ensure the normal development of the trait controlled by the particular gene. Moreover, mapping the human genome will facilitate the overall ability of parents to make the personal eugenic decisions that they have been making anyway, throughout the whole of human history. On the other hand, there are legitimate concerns that the technology will result in a "parade of horribles": invasions of privacy—such as the use of genetic information to affect employment, medical insurance, or life insurance—or, perhaps, efforts to eliminate certain genes from the human gene pool. There are other moral hazards as well, including attempts by government or private corporations to gain proprietary interests in the human genome.

Some of the hazards have already begun to emerge. For example, several years ago, the United States government asserted that it would pursue patent rights on those portions of the human genome that described how the brain functions.

Although the government abandoned this effort, the matter of proprietary interests in the human genome remains to be resolved. More recently, efforts to isolate the genome of particular ethnic groups have met with understandable skepticism by Native American, Australian aboriginal, and other indigenous groups, who are concerned about such issues as the use of resulting information to undermine land claims or to create biological weapons customized for use against these groups.

Martine Rothblatt's bold and pathbreaking book is designed to explore the tension between the possible benefits and hazards of this emerging genetic technology. Rothblatt's goal is to ensure that we as individuals achieve the "genomic literacy" necessary to make intelligent and morally sound decisions about how to use this technology. She accepts that "history is replete with human efforts to at least manage, if not control, the biotechnology of reproduction. It is apparently an unshakable part of human nature to want some control over the characteristics of our offspring." Although Rothblatt qualifiedly regards personal eugenics as a fundamental human right, she distinguishes between parental decisions to influence the creation of new genomes and governmental and corporate policies that seek to control those private eugenic decisions. Because the latter represents the repression of the former, Rothblatt argues, she is highly critical of social eugenics.

Rothblatt proposes that what she calls the "bioethics of birth" be guided by notions that favor individual freedom and reject discrimination. She identifies principles that seek to ensure that the human genome is regarded as the common heritage of all people and that personal genomic choices are protected and facilitated as a matter of social policy, and she advocates a prohibition on governmental influence over genomc expression and a prohibition on any form of genomic discrimination. Rothblatt emphasizes and demonstrates convincingly that the principles that she proposes, as well as her overall framework for regulation of genomic expression, fit very comfortably into our already accepted social, political, and legal norms.

Further, Rothblatt recognizes that distinguishing between personal and social eugenics is complicated by discrimination based on sex, race, and class. Accordingly, her bioethics of birth addresses the need to educate people about the continuum of genomic characteristics, offer public assistance to ensure that all people can exercise their procreative rights, and guarantee genomic health care services to all to ensure that there is no genomic oppression of disempowered groups.

Rothblatt's book also presents fascinating—and admittedly controversial—means for implementing her bioethics of birth. She introduces a concept called "inocuseeding," which might involve banking all men's semen and performing a vasectomy on each. The banked sperm could then be used to fertilize eggs when desired. Inocuseeding would ensure voluntary genomic creation, in that becoming a parent would involve a deliberate decision rather than an unintentional occurrence, as is the case in approximately 50 percent of births at present. Rothblatt regards unwanted pregnancies as a form of disease, and she urges free access to abortion to ensure intended and voluntary genomic expression and to remove human reproduction from the patriarchal framework in which it now exists.

Rothblatt's book—like all books that seek to shift our paradigms of thought in fundamental ways—raises as many questions as it answers. For example, although Rothblatt does not want the government to be involved in genomic decisions, her bioethics of birth requires significant governmental involvement in prohibiting discrimination against genomic expression and ensuring that each person be able to exercise his or her right to engage in intentional genomic expression. As Rothblatt recognizes, it may be difficult to limit the role of government in matters genomic to that of a benign facilitator of personal decisions. Her concept of inocuseeding raises many issues, including the danger that governments or other unfriendly interests may find it easier to destroy the genes of unpopular people or groups through the destruction of sperm banks in which the sperm is kept. Finally, Rothblatt's (again qualified) support of transgenic creations raises issues about whether humans have the right to disrespect the genomes of nonhumans.

Whatever the reader may think about any particular idea contained in *Unzipped Genes,* one thing is certain: The ability of humans to intervene in personal genomic expression is on the verge of entering a phase the likes of which have not been seen in human history. Within a short period of time, the Human Genome Project will map our chromosomes and gain an understanding of their function. The issue is not *whether* we need a bioethics of birth. We *necessarily* need one. Martine Rothblatt's proposal is both thoughtful and provocative and will most certainly be a significant contribution to our social discourse about these fundamental issues.

GARY L. FRANCIONE

author's preface

Throughout history, we have never had to worry about a bioethics of birth because, well, babies just happened. Bioethical concerns were pretty much limited to *man*kind's efforts to ensure that the children their women birthed came from *their* intercourse, and not that of another man. But the cloning of sheep and monkeys shows things are changing very fast. Governments around the world are massively funding a Human Genome Project that will completely map our human chromosomes just around the year 2000. Soon thereafter, before a child is conceived, the parents will be able to know what the child will look like, which diseases the child will be predisposed to get, and how the child may behave. And then will come the big question: Is there anything you would like us to change?

This book outlines the bioethics of birth that will prepare us for the human genome questions of tomorrow. The bioethics of birth are derived from the principles that have served us well over the past two hundred years: freedom of expression, prohibition of discrimination, and scientific public health. These time-honored and frequently refurbished principles must do duty for us once more, this time in the brave new world of genetic engineering. They are needed to ensure that unintended births will not be forced upon us, that intended births will not be denied us, and that the very question of what is or is not an intentional birth will be left to individual conscience and not government decree.

The bioethics of birth, applied promptly and properly worldwide, will ensure that the benefits of biotechnology will not be lost in demographic holocausts of one or another genetic trait. Eliminating a genetic trait from the human genome can mean eliminating a demographic group from future generations of human life. This book welcomes the brave new technology of intentional evolution, but attempts to guide it in the ways of our somewhat civilized world. I am a biotechnology convert, but not a mindless defender of genetic free-for-alls. We cannot survive without biotechnology and we cannot survive its unethical use. Indeed, we have no choice but to develop a bioethics of birth. The only question is how many demographic deaths will be suffered before we make the right choices.

In order to make right choices, people need pertinent and reliable information. Unlike the highly specialized discourse of other learned treatises on the subjects of genetics, eugenics, bioethics, or biotechnology, this book provides information in terms that are accessible for all of us.

One of the genetic entrepreneurs whom I interviewed for this book was Dr. Craig Venter, chief executive of The Institute for Genomic Research and inventor of the computer-scanning technique that made decoding the entire human genetic code practical. I asked him what was needed most to ensure that the Human Genome Project's knowledge would be used for good rather than for harm. His response was simple: "genomic literacy."

At first, I was skeptical that the American masses, not to mention the world's masses, would understand the intricate genomic lingo; I could barely pronounce the vocabulary (deoxycytidine monophosphate), scarcely recognize the research goals (alpha-1-antitrypsin), and hardly solve the basic equations ($H^2 = (r_M - r_D)(1 - r_D)^{-1}$). Was it realistic to expect that the general population would become genomically literate, as

Dr. Venter had desired? (The equation gives the likelihood from 0 to 100 percent of the heritability of a particular trait; the research goal is a genetically coded protein the absence of which promotes cirrhosis and emphysema; and the vocabulary word is one of the four molecular building blocks of DNA.)

In talking about fetal-testing technology with Asian women, speaking on bioethics with American students, and discussing eugenics with European activists, I found that the goal of genomic literacy was actually quite achievable, indeed. One does not need to spell out DNA to understand that inside every part of one's body is a microscopic chemistry instruction book for turning food and air into flesh and blood. (How else could the body be built and maintained?) One does not need to know the fancy names for all our proteins to realize that each person's instruction book is a little bit different. (That's why similar habits do not necessarily result in similar diseases.) And one does not need to know any algebra at all to deduce that half of our personalized biochemical instructions come from a sperm donor and half from an egg donor. Most of these instructions are identical, so we pretty much have two copies. But some are a bit different—and they make us different. Sometimes these instructions get garbled, and the result can be what we call sickness or disease. Sometimes these instructions are changed intentionally through selective breeding or test-tube tinkering. That's what we call "eugenics." And sometimes all the similar instruction books are burned. That's what we call a holocaust or demographic death. In short, the basics of genomics are understandable by anyone.

I found that nearly all published works on genetics contain roadblocks to mass genomic literacy in the form of scary-looking diagrams of molecules, endless strings of footnotes, and lots of specialized discourse about recessive and dominant traits. Of course, mass genomic literacy was not the objective of these treatises; they were written for biologists, sociological demographers, and legal scholars. But the future of our human genome is far too important to be left to specialists.

It is essential that we as citizens be able to dispute laws that would allow some people's microscopic instruction books to be passed on to their offspring, while consigning other people's genetic heritage to be interred with their corpses. It is critical that people be able to cry foul when authorities claim that scanning their genome for psychological traits is no different from scanning their blood for psychoactive drugs. And it is only fair for all people to participate in establishing our nation's basic policies of birth, without first being obliged, as if in some modern-day poll test, to provide a second-by-second biochemical account of how zygotes are formed.

Just as verbal literacy is essential to protecting democracy, genomic literacy is essential to protecting demography. We do not need to be constitutional scholars to separate political wheat from political chaff, and we do not need to be biochemical scientists to separate bioethical truth from bioethical fraud. All that we need is a chance to learn the basic rules of the game. These rules are clearly laid out in the chapters that follow.

This book is a call to action, a manifesto for the adoption of a new bioethics of birth. We must formulate our genomic ethics now, or we will assuredly suffer from its absence later. The time has come to declare, for the benefit of all persons living and those not yet born, that demography is as important as democracy, and that an attack on any one demographic group is an attack on us all. The time has come to snuff out the possibility of genomically inflicted demographic death, and to usher in the new millennium with a strong and protective bioethics of birth.

acknowledgments

First and foremost, I want to thank Andrew Fisher and Paul Mahon for their great efforts as agents, editors, and reliable friends. This book owes itself to their encouragement, advice, and marketing acumen. The humor, wisdom, and companionship of these guys, and their associates Chris and Jerry, are invaluable assets to me on the journey through life.

The editors at Temple, especially Jennifer French, Mary Capouya, Joan Vidal, and Doris Braendel, and Series Editor Gary Francione, have been wonderful. Every writer should have the opportunity to benefit from the understanding of semantic nuance and clarity of style that comes so naturally to Ms. Capouya and Ms. Braendel.

I owe a great intellectual debt to Dorothy Roberts, Susan Wolf, and the Center for Responsible Genetics. Susan Wolf clarified for me the similarity between seeing people as genes and seeing people as sexes or skin tones, and especially the fallacy of thinking that antidiscrimination laws cure discrimination. The laws help, but it is the mind-set of seeing people as biology that must be changed. We are all part of a continuum of humanity. While I chose to use the term "genism" rather than her term "geneticism," I believe the intent is the same and the credit is hers. "Genism" has the benefit of allowing us to flag the bad guys as "genists" rather than as "geneticists" (*smile*). Dorothy Roberts is responsible for letting the world see that simply grafting genetic technology onto a racist and classist social structure will only reinforce existing oppression. I thank her for her brilliant essay, "The Genetic Tie," and for the most trenchant and thoughtful critiques of my early manuscript.

The Center for Responsible Genetics, in Cambridge, Massachusetts, is humanity's watchdog agency against the hype and hubris of the genetics industry. I have learned much from its members, especially at a roundtable discussion they organized at my request, and from personal talks with its founder Ruth Hubbard and Executive Director Wendy McGoodwin. While the center and I do not agree on all things genomic, I believe that there is no organization more deserving of moral and financial support in the grassroots battle to avoid genist oppression.

I have an enduring sense of gratitude to my superb college professors, especially Patrice French, who taught me the value of out-of-the-box thinking; Paul Rosenthal, who taught me how to reason most persuasively; and Charles Firestone, who always encouraged my scholarly research. And I will never forget my first multicultural studies teacher, Mr. Kiriyama of Louis Pasteur Junior High, who spent World War II as a dispossessed person behind a barbwire fence in a desert internment camp for innocent Japanese American civilians. He taught us well that we are so much more what the world makes of our genes and our culture than anything intrinsic to those genes or culture. This became the seed of my fervent belief in the ever so much greater power of euthenics over eugenics.

I am also honored to acknowledge the historical contribution Madame Noëlle Lenoir has made to this field through her leadership of UNESCO's International Bioethics Committee (IBC). Her great accomplishment is in almost single-handedly raising global awareness of the human rights dimensions of genomic technology. This book benefits immensely from my careful study of the deliberations of her committee, and I applaud the openness and inclusiveness with which she has conducted her work.

The International Bar Association (IBA) has similarly made a historical contribution through its sponsorship of the first-ever international treaty on the human genome, developed over the period from 1993 to 1996. The IBA, representing some 183,000 lawyers worldwide, has accomplished a singularly effective mission by bringing the need for genomic legislation directly to the attention of all the world's leading legal authorities. I thank Tom Forbes, Leigh Middleditch, Jr., and especially Paul Honigmann for their consistent support of my efforts as vice-chair, co-chair, and finally chair of the Bioethics Subcommittee of the International Bar Association. When a Human Genome Treaty finally becomes law, much of the original credit for its content will go to Paul Honigmann, whose early drafts of the treaty met with much acceptance worldwide.

The IBA and IBC appear to have simultaneously introduced the concept of the human genome as the common heritage of humanity, the IBA at the first meeting of its Bioethics Subcommittee, in October 1993 in New Orleans, and the IBC at its first meeting that same month in Paris. The IBA came to this view from a resource law perspective (outer space and the high seas are "common heritage" resources), while the IBC came to this view from a cultural heritage perspective (art and culture are considered the "common heritage" of humanity).

I am grateful to Mr. Noah Azmi Samara, chairman of WorldSpace Corporation, for his sponsorship of my travel to the IBA bioethics meetings, and for my opportunity to promote the creation of such a revolutionary new common heritage legal concept. I also greatly appreciate the expert research and indexing assistance of Teresa Bongartz.

My greatest acknowledgment is to my spice and soulmate, Bina Aspen Rothblatt, without whose unflagging encouragement this book would never have seen the light of day. She never wavered in her support for the themes and messages of this book. Indeed, she never rested in demanding that I finish the book for the benefit of people everywhere, but especially for the women of tomorrow.

part I sex and the genie of life

Every individual is a universe carrying the
memory of humanity in their genes.

PAUL ASTER

"Genome" is the modern word for a unique combination of deoxyribonucleic acid, or DNA; for humans, it is a complete set of genetic instructions spread across 23 pairs of chromosomes. Half of a human genome's DNA comes from the head of a testicle-spawned seed called a spermatozoon ("sperm") and the other half comes from the body of an ovary-hatched seed called an ovum ("egg"). When these two seeds merge they create a unique genetic combination, a new genome, contained within a hybrid sperm–egg pair called a "zygote." With some luck, the zygote may self-replicate and grow so that the genome eventually expresses itself as the body of a living, breathing person. But the genome will never actually be "the person," just as the acorn is not the oak tree. People are experiences, the endlessly unique and often chance results of bodies (including the part between our ears) sensing the surrounding world, and the surrounding world changing those bodies (and minds).

We can speak about either a "person's genome" or the "human genome." A person's genome is unique to that person, an identical twin of that person, or clone, if any. It is very specific and concrete. My genome is the entire set of genetic instructions contained within my own special form of the DNA molecule, inherited at conception from the seeds of my parents' sex. The human genome, on the other hand, consists of the genetic instructions common to all people's genomes, plus all the knowable variations. The human genome is an abstract concept because there is no such thing as a DNA molecule that contains genetic instructions for every different kind of person. A DNA molecule can describe only one person's genome at a time. The human genome is a mammoth catalog; a person's genome is one lengthy order out of that catalog.

By way of example, each person's genome tells that body how to create that heart. Of course, hearts come in many different sizes and abilities to pump blood. Some hearts are born with missing pieces. The human genome describes all the different ways that a body can create a heart. Clearly, the part of my genome describing how to make my heart is very real and specific. On the other hand, the part of the human genome dealing with the heart is a set of many different possibilities.

We are not our genome, or its DNA, although it is very much a part of us. Our genome is like the car that takes us on the journey of life. But it is no more the journey than the family car is the summer vacation. Genomes, like cars, are tools. People are the result of using those tools in the world around us. Good care of a "bad heart" may take a person a lot farther in life than bad care of a "good heart." On the other hand, the nature and soul of a person have nothing whatsoever to do with the mechanics of the person's heart.

Genomes can come from "bed sex" or "lab sex." Bed sex, to yield a new genome, means a couple in bed with penis and vagina fully engaged. For the genome, though, lab sex will work just as well: In vitro fertilization and artificial insemination also produce new genomes. There is nothing perjorative about lab sex. After all, the only reason sex exists at all, say the experts, is to ensure genetic diversity and hence survival of the species. One serves Nature's will just as assuredly with seed-creating techniques that are artifactual (turkey basters, test tubes) as with the coital. In any event, an artificially produced genome is not an artificially produced person. People do not differ based on how their seeds were sown.

Biotech companies are revolutionizing our ability to know and affect the seeds of sex. Advanced genetic technology promises to describe with great accuracy the future appearance and likely predispositions of human life. This is done by scanning a genome for specific DNA sequences that always mean the same thing in the experts' increasingly comprehensive gene maps. Some DNA sequences may tell "eye color," others may signal "obesity." Some genes may signal "future breast cancer," while others may assure "perfect pitch," "heterosexuality," or "depression."

Once again, however, a person is not an eye color, medical condition, or even a mental predisposition. People are totally unique sets of experiences out of an infinite variety of possibilities. While the life-sensing tools represented by a person's genome—say perfect musical pitch—may direct the individual toward some rather than other experiences, this fact in no way limits the unique personhood of the individual. Numerous other factors have even greater effects on our set of possible experiences (where, when, and with how much money, love, and encouragement we are raised), and humans are ingeniously creative creatures. From almost any set of genetic tools, humans have an uncanny ability to achieve an unpredictably diverse array of experiences. Our greatest space explorer, Stephen Hawking, fights his own nerves and muscles just to move. Yet many 6-foot-tall men, with 20/20 eyesight, and 130 IQs, spend most of their lives in office buildings shuffling paper. Mapping a genome, and even predicting the corpus it can become, is still in no way mapping a person and predicting the kind of life that person will live. People and lives are experiences, not molecules.

We carry our genomes with us from life after sex to life before death. Hence the genome scanners can read our future propensities when we are brand-new 8-cell creatures as small as a needle tip, or much later when we decide to become parents, or later still when we face the disease lottery of old age. Indeed, since science's ability to decipher the genetic code seems to improve every month, it is likely that our genome will soon be scanned regularly, from life after sex to life before death.

For children or adults, the new genomic technology can be a genie of life. Early diagnosis of diseases, or propensities to disease, often result in fantastically successful cures. But for the seeds of sex, the new genomic technology is usually used as a genie of death—a tool to selectively target for extinction any genome that has unpopular traits. As will be shown later, the first large-scale use of genomic technology has been to inflict mass demographic death on Asian women. When routine pregnancy scans in much of Asia reveal a female fetus, an abortion is usually performed. In India and China particularly, the young female population is dropping precipitously below that of the male. It is not popular to be born with a vagina.

A November 30, 1993, *New York Times* article reported that 11 percent of Americans would abort a fetus whose genome scan predicted obesity. About four out of five people would abort a fetus if it would grow up disabled. And 43 percent of respondents to a March of Dimes poll said they would engage in genetic engineering simply to enhance their children's looks or intelligence. For the seeds of sex, genomic technology will often be a genie of death.

Unfortunately, many people are confusing the genome with the person. They are judging a soul by its wrapper. They are trying to give their offspring a "better vacation" by giving them a "better car." But can we blame them? Few among us would prefer "worse" tools to "better" ones. The problem is one of focus. The more we focus on the body, the genome, the more we neglect the overarching importance of human experience. At some point, we will have poured most of our resources and admiration into the new car and will no longer be able to enjoy the vacation. If we "must enjoy the vacation" because of the "superb car," then it is pretty much preordained that we'll hate the vacation. Nevertheless, it is inescapably human to "want it all." Most people will believe that they can shape their children's genes—such as by aborting fetuses a genome scan predicts will grow up disabled—and still offer their children wonderful life experiences.

Danger arises when an outside force, especially a governmental authority but also simple social pressure, tells us which seeds to discard and which ones to keep. One problem here is is the tremendous disagreement over what constitutes a disease, disability, or disorder. One man's disease may be another man's inspiration. One woman's *dis*ability may be another woman's *different* ability. One person's disorder may simply be another person's personality. When others tell us what kind of genome to create or discard, they have made a fatal leap from the genome as a tool set to the genome as a person. They will have said "genomes with that characteristic cannot exist." This is quite different from individuals deciding to abort based on genomics, for the individuals reach the decision only for themselves. Individuals are not denying genomes with any particular characteristic the right to exist at all. They have unfortunately confused the genome and the person, but they have affected only their own offspring. When the government or mass ethics confuse the genome and the person, all people with the genomic characteristic are affected.

When certain genomes are not allowed to exist, demographic death will occur. A trunk of human life will become extinct, its unique genomic contribution forever lost. Worse still, persons still living with the proscribed genomic characteristic will be seriously devalued—not even deemed worthy of reproduction—often leading to a loss of their civil rights. After deciding which groups of people were "inferior," the Nazis first sterilized them, then interned them, and finally killed them. Each step facilitated the next.

Perhaps the final horror of genomic selectivism is that it invariably acts like a relentless pathogen. The death of one demographic group makes it so much easier to justify the extinction of other genomic traits. Once unleashed, the genie of death may become unstoppable. Setting social groups against one another based on their genomic inheritance will reverse the current trend toward multicultural global unity with a regression into biological tribalism. In brief, compulsory genomic selectivism is a trigger for warfare, a trigger we should do our utmost to remove, a genie we should keep bottled up for good.

Whether the new genomic technology becomes a genie of life or a harbinger of death will depend on our ability to rapidly implement a new bioethics of birth premised on four key principles:

Absolute prevention of discrimination based on a person's genome; coupled with education on our shared continuum of genomic characteristics;

Complete freedom to parent any genomes we please, coupled with society's obligation to outreach aggressively to its nonmainstream components and assist them in exercising procreative rights;

Total prohibition of any governmental influence over genomic expression, coupled with a governmental obligation to ensure that de facto class-based genomic oppression does not occur, via the guarantee of genomic health care services to all;

Universal respect for the human genome as the common heritage of all humanity.

These principles all combine to make sure that whatever our seeds of sex grow up to look or act like, they will be accorded the same respect as every other human being. The bioethics of birth will accomplish this singular objective for a new genome in four complementary ways: (1) vesting all of humanity with an interest in its existence, (2) depriving governments of the power to legislate away its demographic, (3) ensuring its right to reproduce, and (4) making its socioeconomic life viable.

We are now ready to begin our journey through the mysteries and intrigues of the seedling phase of life after sex. It is a journey that will take us through Third World villages and First World laboratories, and from the precipice of neo-Nazi population policies to the biotech vanguard of personal liberty. The journey will culminate with creative new solutions to the "choice versus life" abortion controversy and with a solid new legal foundation for maintaining respect for individual diversity in an age of genomic predictability.

chapter 1 the holocaust of sex

The first fetal condition that could be diagnosed prenatally was, ironically enough, sex.

RUTH SCHWARTZ COWAN

1992

Around 1990, demographers working on the United Nations Educational, Scientific and Cultural Organization's (UNESCO) annual statistical reports of birthrates first began to notice that the number of female births in India and China was dropping far below that of males. Through the 1960s and 1970s, there were about 106 boys born for every 100 girls. But by 1984, the scholarly *Population and Development Review* reported that second-child ratios had leapt to 114 boys born for every 100 girls, and by 1989 to 121 boys for every 100 girls. The Chinese and Indian governments' own census statistics show yet more skewed ratios for third and fourth children of over 130 boys for every 100 girls. A February 14, 1993, *Washington Post* article reported that these statistics added up to "77 million missing Asian girls—more than the entire combined populations of California, New York, Texas and Florida." This veritable holocaust of sex was noticed in village schoolrooms: There were ever fewer girls.

Throughout history, the number of males born has slightly exceeded that of females. The slightly higher natural male birth results from the biological difference between males and females in the 23rd pair of their chromosomes. Males ordinarily have a so-called X and Y pair, whereas females usually have an identical XX pair (named for the approximate shape of the chromosomes). Sperm and egg cells, unlike all other cells, carry just a single strand of 23 chromosomes. With women, all of their chromosomes are identical pairs, so it doesn't matter whether a particular egg cell carries one or the other of a woman's double X chromosomes. But since a man's 23rd chromosome is XY, half the sperm cells have an X and half the sperm cells have a Y. For some as yet unknown biological reason, sperms carrying a Y chromosome have a slightly higher probability—about 6 percent—of successfully fertilizing an egg cell than do sperms carrying an X chromosome. This difference is responsible for the slightly higher percentage of male than female fetuses—the historical 106 boys born for every 100 girls noted above.

In most countries, the sex imbalance usually evens itself out because males die at a higher rate than do females. This is due to the greater percentage of males who suffer death due to disease, accidents, and violence. However, in a few developing countries, the original imbalance is widening because of the deliberate withholding of health care and adequate nutrition from females, the health risks associated with pregnancy, and male-on-female violence.

The social researchers who discovered the large and unusual difference in birth rates between males and females had inadvertently stumbled upon the largest effort ever to suppress the birth of a particular demographic group. Going into the field, demographers quickly learned the reason for falling female birth rates. In the mid-1980s, American companies had licensed ultrasound and amniocentesis technology to the Asian market. The ultrasound technology permits a trained operator to discover whether a fetus has a penis or a vagina, although errors are common. Amniocentesis, first demonstrated in 1955, is used to discover with high accuracy whether the fetus has an XX or XY chromosome structure.

Local Indian and Chinese firms quickly geared up to manufacture the sex-determination biotechnology. Using low-cost local electronic components, ultrasound machines were produced with a selling price under $2,000. Amniocentesis laboratory test equipment was made available in the Indian market for less than $3,000.

Distributors for the Indian and Chinese manufacturers contacted doctors, hospitals, and clinics in large cities and rural areas. They explained the technology. Little needed to be said about the economics of its use. It was obvious to any medical professional that people would pay a hefty fee to learn the sex of an embryo in time for a female embryo to be safely aborted. By 1990, well over a hundred thousand units of the sex-determination technology had been sold.

The reasons for the high consumer demand for fetal sex information varied from place to place. When the technology had initially become available in the United States and Europe, it was ostensibly to be used to scan the fetus for signs of future medical problems such as Down's syndrome or spina bifida. Information on the fetus's sex was given largely to satisfy the curiosity of expectant parents. Should the baby's room be painted pink or blue? Gender training is considered one of parents' biggest responsibilities, and many parents wanted a headstart on this brain-molding task.

The sex-determination technology could also be used to

fine-tune the likelihood of a child's being born with a debilitating hereditary disease. For example, if a parent carries a Y-chromosome-linked hereditary disease, such as hemophilia, the child of unknown sex has nearly a 50–50 chance of getting it. A female baby will rarely be born with it. Hence, some parents with a family history of genetic conditions such as hemophilia were counseled to consider aborting a male fetus rather than risk having a congenitally ill child.

There is also a lot of anecdotal evidence that many Western parents used the sex-determination technology to achieve a desired mix of male and female offspring, aborting any fetuses of the undesired sex. As the science historian Ruth Schwartz Cowan has observed in her essay, "Genetic Technology and Reproductive Choice," by the mid-1970s, "word eventually leaked out" that amniocentesis intended for predictive medical genetics could be used by a woman "to discover the sex of her fetus in time to have a legal abortion if she was not happy with the sex of the fetus she was carrying." Cowan notes that while many American physicians felt that sex-selective abortions were unethical, patients were beginning to request amniocentesis "solely for the purpose of discovering the sex of their fetuses, most frequently women who had already had a large number of children of one sex." No demographic anomalies developed because, in late-twentieth century Western culture, an ideal family is based on an equal sex ratio among offspring.

In most of Asia, female child abortion is the reason behind sex-determination technology's popularity. There is a prevailing ethic in most of Asia that male offspring are preferable to females. China's "one family–one child" birth control policy means parents must abort female fetuses if they wish to achieve their male offspring preference at all. India's female dowry custom is so expensive that most families have little choice between aborting female fetuses or family impoverishment. Korean families are reluctant to waste scarce resources on a female child's education when widespread sexism will bar her from ever achieving as rewarding a career as a man.

The practice of female fetal abortion has ancient and widespread roots in female infanticide. Indeed, the killing of newborn baby girls continues today in some parts of rural South Asia. Some selective abortion defenders argue that female fetuscide is a more ethical alternative to female infanticide. There is no doubt that female fetuscide is easier to accomplish than infanticide. It is, of course, always especially gut-wrenching to take a living, breathing infant to a nearby field and suffocate it under a mound of dirt. A widely practiced alternative involves drowning the infant in scalding water immediately after birth. In some locales the infant is killed by forcing poisons down her throat.

In contrast, a visit to the sex-determination and abortion clinic is a largely aseptic event. Often sex determination and abortion are accomplished in two different clinics. This permits everyone involved to plausibly deny that they are aborting the child because it is female. Unfortunately, as noted below, it is unlikely that sex-selective abortion replaces female infanticide; instead, they supplement each other.

In the near term, it is likely that Asia's holocaust of sex, left unstopped, will slow down or perhaps reverse course, because women as their numbers diminish, will become more valuable. Economics would suggest that dowries would begin to be paid *to* the parent of girls, and parents would then be more motivated to abort male rather than female fetuses. But such a reversal is probably many years off. A May 11, 1996, *Washington Post* front-page story reported:

> [The newest] research on demographic data from India, China and Taiwan suggests that biases against girls are not disappearing even as those countries' birth rates approach those of Western nations. The work also indicates that women are increasingly using pre-natal sex-determination tests and selective abortions to ensure that they give birth to boys. . . . [S]elective abortion is not substituting for female infanticide [a poor person's option], but supplementing it.

At the 1996 annual Population Association Conference, Harvard researcher Monica Das Gupta glumly concluded that females missing due to neonatal neglect, outright infanticide, and selective abortion are "additive" and that by "the year 2000 we should expect to see more of it."

Notwithstanding the worsening trends, the greatest danger from this partial holocaust of sex is not the total elimination of women. Socioeconomic forces are likely to prevent a total holocaust. Aside from the horror of a partial holocaust, the greater danger is a bioethic of birth that permits selective termination of life based on its demographic characteristics. Such a bioethic creates a slippery slope along which all manner of fetuses or genomes are terminated based on chromosomal analysis of their likely appearance, attitudes, or behaviors. On one hand, this leads to a devaluing of the worth of those living persons whose characteristics can become triggers for abortion. More perniciously, the slippery slope leads to demographic death for entire categories of human beings.

The principal reason women agree to abort female fetuses is that they realize life will not be pleasant for women in a patriarchal society. In India, for example, women and girls traditionally cannot even start to eat until the men and boys have had their fill. In tough times, this means hunger pangs for the women and consequent susceptibility to malnutrition and disease. If a daughter falls ill, women know that they will not be allowed to spend money for her medical care—that is reserved for the boys. Female infanticide is the result of a mother's nurturing instinct telling her that, due to the sexism of society, the girl is better dead than bred.

Yet, is life any better for the handicapped in an ablist society, the gay in a straight world, or the obese in Barbiedom? The technical possibility for demographic death of each of these groups, and a great many others, exists as a consequence of an unprecedented international endeavor—the Human Genome Project. This project, discussed in Chapter 3, will result in the identification of every unique genetic "marker" that is likely to signify membership in one or another demographic group. The "markers" will be scannable in any single cell, even in a seed of sex genome. Hence, the Human Genome Project gives rise to the ability to abort any fetus based on its predicted demographic characteristics.

The Human Genome Project may inflict its own demographic Hiroshimas before being disciplined for peace. Indeed, as shown above, simple X versus Y genome technology has already been used demographically to zap some 77 million female seeds of sex. But we *can* still stop this democidal juggernaut and avoid repeating the most horrible aspects of our history. We can marshal the world's collective foresight and willpower and adopt a mandatory bioethics of birth before we suffer from the debilitating demographics of death. To do so, we must be sure to understand clearly what the Human Genome Project is, where it is heading, and who controls it.

chapter 2 genomic imperialism and democide: the age of discovery reborn

We have to reassure people that their own DNA is private and that no one else can get at it.

JAMES WATSON
human genome czar, *1989*

The genetic diversity of people now living harbors the clues to the evolution of our species.

LUIGI LUCA CAVALLI-SFORZA
global genome collector, *1991*

A human genome is a person's entire set of chromosomes. There is not a single human genome because no two people, except identical twins, have identical chromosomes. However, about 99.99 percent of human chromosomes are the same for all people. The small differences account for much of our physical uniqueness from one another. The term "the human genome" means the 99.99 percent of our chromosomes that we all have in common, plus all of the knowable variations in the other .01 percent.

There are usually 23 pairs of chromosomes in a genome, although some people have more than two of a chromosome. This means that people have 46 or 47 chromosomes in total.

Each chromosome is a long molecule called DNA and made up of a pattern of short molecules—like a homemade bead necklace. If the short molecules were the size of typical beads, each chromosome would be miles long. Instead, 100 short molecules are only one-millionth of an inch long. So the entire genomic necklace fits easily inside a human cell.

Each pair of chromosomes is usually wrapped round and round one another like the longest spiral staircase imaginable. A rough idea of a chromosome's shape can be visualized by taking a bead necklace and twisting it from each end. The clumpy spiral pattern that results is how each chromosomal pair looks when packed away inside a cell's membrane.

Each human cell, except for eggs and sperms, has an entire genome. Hence there are billions of genomes in our body—one for each cell. The egg and sperm cells each contain only half a genome. Scientists call egg and sperm cells "germ cells" because of their ability to *germ*inate new life. Females usually are born with about a million germ cells (eggs) and gradually lose them over a lifespan. Usually, males first begin producing germ cells (sperm) at puberty, creating millions each day. The ability to continue producing germ cells ebbs as the male ages.

Each of the 23 different pairs of chromosomes has a unique pattern of short molecules. The pattern shows up under ultraviolet light and a microscope. These patterns let scientists separate the chromosomes by number. A particular chromosome always follows the general pattern shown in Figure 1, whether it is in one of your eye cells or toenail cells, or whether it is in one of your best friend's lung cells or sperm cells. The patterns only show up clearly if the chromosomes are first made to unwind and lie out flat. When clumped up like spiraled beads it is almost impossible to tell chromosomes apart.

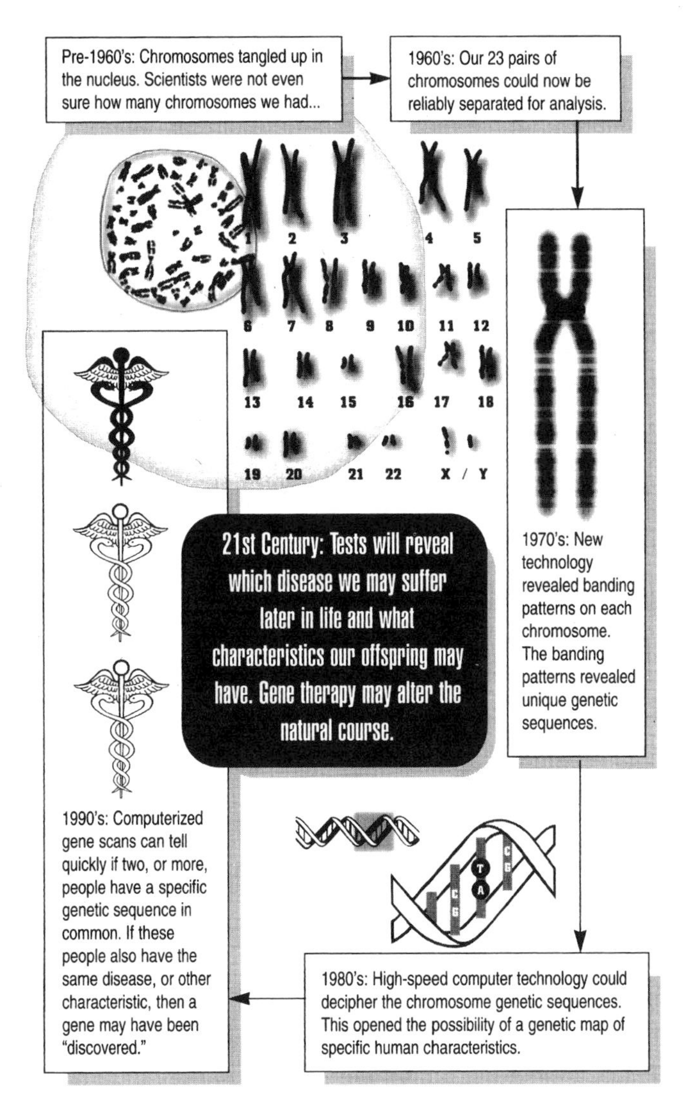

FIGURE 1: The Age of Genomic Discovery.

Source: Martine Rothblatt

It has only been in the last thirty years that scientists have learned how to make chromosomes lie out flat and reveal their small genetic "bead" patterns.

The patterns on chromosomes reveal the short molecules of which they are made. Each short molecule is like a few beads on a necklace. Each has the ability to instruct nutrients in the body to form one of the thousands of different proteins and fats that make the body work and look as it does. It is best to imagine each short molecule as a kind of magical magnetic cookie cutter. The magic arises because putting the beads together in just such an order creates a magnetic cookie cutter instead of just a string of beads. This is like saying that words in just the right order "casts a magic spell." Any other order of words is just a string of words.

Once our DNA has created a magical magnetic cookie cutter, the baking begins. The cookie cutter magnetically attracts to it from surrounding cells, blood, and tissue the nutrients it needs to make a specific building block of the body. And then it shapes those building blocks the way a cookie cutter shapes a cookie. The building blocks are "shaken out" and ejected when done. The molecular cookie cutter immediately starts to attract new nutrients from which to bake another building block, while the fresh-baked building blocks float away to do their job within the body.

Each string of short molecules—each magnetic cookie cutter—has the ability to instruct the formation of only one kind of nutrient combination (body building block). Thus it follows that there must be a great many different kinds of short molecule strings. After all, it must take many different kinds of building blocks to make everything from blood to hair to bones. In fact, it takes about 100,000 different kinds of building blocks—

arranged in many more different combinations—to make a human being. Thus there are about 100,000 different strings of short molecules spread across our 23 pairs of chromosomes.

Since about 1900, scientists have called these strings of short molecules "genes." Each unique string or magnetic cookie cutter is a unique gene. Hence, the human genome consists of about 100,000 genes, each of which instructs the body to do some particular microscopic building-block task. Taken together, these thousands of genes form 23 unique patterns, one for each chromosome. It is especially remarkable to realize that our 100,000 magical magnetic cookie cutters are at work 24 hours a day, side-by-side, inside each of the billions of cells of our body. It remains largely a mystery as to which cookie cutters are active within which cells and exactly what paces their work.

The greatest miracle of life is how the 100,000 genes work together in such harmony that their output creates a single, highly functional living being. The columns below summarize some of the metaphors we have used in this section, to help visualize exactly what kind of extraordinary gift we are beginning to tinker with in the Human Genome Project.

Reality	*Metaphor* 1	*Metaphor* 2
Gene	Unique bead sequence	Unique cookie cutter
Chromosome	A necklace	A drawer of molds
Genome	23 necklaces	23 drawers of molds
Cell	Jewelry box	Magical bakery
Body	Billions of jewelry boxes	Billions of bakeries

There are about 3 billion individual short molecules, or "beads," in a genome. About 10 percent of the beads are grouped into recognizable patterns of strings. The rest of the beads show no pattern at all. It's like having a necklace with well-thought out patterns separated by random assortments of beads. Scientists believe the randomly arranged short molecules have no purpose. Hence they call these molecules "junk DNA." The 10 percent of the molecular beads that are grouped into 100,000 subpatterns are thought to contain all the biochemical information necessary for a human being.

Since there are about 3 billion short molecules overall, 90 percent of which are "junk," this means that about 300 million short molecules must be deciphered to know completely how a human being is made from a genome. As noted above, scientists have identified the existence of about 100,000 genetic patterns from these 300 million short molecules. This implies that each genetic pattern contains, on average, about 3,000 short molecules (100,000 times 3,000 equals 300 million).

The scientific mission of discovery is to reveal what the genes within our genome do. To "map the genome" is to know what it does. To isolate and repeatedly identify about 3,000 short molecules out of the seemingly endless DNA, and associate this 3,000 molecule string with some known human phenomenon, is to discover a gene. To do this for all 100,000 genes is to circumnavigate the genome the way Magellan's crew circumnavigated the globe. For most of the twentieth century, this task seemed impossible. The task of mapping the human genome is usually compared to reconstructing a 23-volume encyclopedia, exactly as it was written, from a barrel of 3 billion letters.

During the 1980s, a few scientists first began to believe the human genome was knowable, was discoverable, or, in Columbus's terms, was round. When the human genome was thought of as 3 billion short molecules that had to be decoded, the problem was unsolvable. There was no voyage of discovery that could realistically find the solution to a puzzle that was 3 billion letters long, all in random order without any known key. However, if scientists disregarded the "junk DNA," that which lay beneath the top of the chromosomal ocean, then the voyage of discovery became ten times easier. And if scientists first concentrated on well-known genetic anomalies—ones passed from generation to generation—the problem might be as simple as finding the same sequence of 3,000 short molecules in similarly afflicted people. Computerized chromosome scanners, first invented in the 1980s, made it possible to search rapidly through different peoples' DNA for matched sequences. This is the same kind of technology used to identify criminals. By the end of the 1980s, a group of influential scientists became convinced that they could prove the genome was knowable, or, metaphorically, was round. They approached what passes today for King Ferdinand and Queen Isabella of Spain, the U.S. Congress, and eventually won a grant to prove their theory. The ships of discovery were dubbed the "Human Genome Project."

why a human genome project?

Just as there were several justifications for Columbus's voyage of discovery—an explorer's quest to explore, merchants' quest for wealth, a nation's quest for global hegemony—a similar array of justifications motivated the launch of the Human Genome Project. The various justifications and players interacted among one another until the project took off in 1988.

The "Columbus" here is a group of people, mostly influential biologists, perhaps the best known of whom is James D. Watson, popularizer of the winding stairway (double-helix) shape of the DNA molecule. In March 1986, one of Watson's helmsmen, 1975 Nobel Laureate Renato Dulbecco, used an editorial in the influential magazine *Science* to call upon the United States to infuse the Human Genome Project with the same kind of effort that had "led to the conquest of space." For many biologists, in Europe and Japan as well as in the United States, it is mostly the thrill of exploration and discovery that motivates their involvement in the Human Genome Project. The "why" for them—why map the human genome?—is "because it's there."

The merchants here are biotechnology companies hungry for profits in a highly competitive industry. Just as Mediterranean merchants dreamed of the profits they would make from exclusive silks brought from Asia, their modern-day counterparts dream of the profits they will make from exclusive ownership of new genetic medicines and cures. Major pharmaceutical companies argued that the cure for cancer would come from deciphering the human genome. Now the biotech companies are no more able (or willing) to fund the mapping of the human genome, estimated to cost several billion dollars, than were the merchants of Venice willing to fund Columbus' voyage of discovery. But business justifications figure prominently in the calculus of the ultimate bankrollers—governments—to fund such expensive undertakings. Dreamers alone cannot tap into governmental largesse. Businesspeople often have greater success. Especially when the businesspeople promise silk, cancer cures, and other forms of economic growth.

The imperial competition for hegemony here comes from

both sides of the Atlantic and the Pacific. While the Human Genome Project has been couched as a cooperative international endeavor from its beginning, such a labeling is really misleading. It is somewhat like calling the Age of Discovery "a cooperative endeavor among England, France, and Spain." To be sure, in both instances, there is cooperative sharing of some maps and tools, but the undertaking is fundamentally a competition among the United States, Europe, and Japan for imperial control of the human genome.

The U.S. imperial thrust began in the mid-1980s when Charles DeLisi, head of the Department of Energy (DOE) Office of Health and the Environment, began to wonder if the government's nuclear bomb laboratories could be recycled into genetic engineering factories. The days of nuclear hegemony were clearly numbered since the Soviets and Americans had begun negotiating reductions in their missile arsenals. But the Americans had built up a vast library of information about DNA mutations, in the expectation of nuclear warfare, and such information was stored at a DOE laboratory in Los Alamos, New Mexico. The computers used for calculating the effects of a nuclear bomb were the closest things around to the kinds of computers needed for deciphering a 3-billion-molecule genetic code. Was it not possible that a new imperial America could reign supreme over the human genome? The international economic rewards from the control of human life could swamp the combined economic value of all the industries America had recently ceded to Asian competitors.

The DOE found ready supporters in the U.S. Congress, as did the scientist–explorers and would-be biotech billionaires. In 1987, Congress kicked off the genome race with nearly $25 million to ensure early U.S. leadership. In the 1990s, this funding had climbed to well over $100 million annually, and the Human Genome Project, now an official federal program, was considered, in the words of genetics chronicler Daniel Kevles, "essential to national prowess in world biotechnology, especially if the United States expected to remain competitive with the Japanese." The age of genomic imperialism had begun.

Within a year of that U.S. effort, similar national efforts were under way in most European countries and in Japan. Although a kind of genomic United Nations was formed, called the HUman Genome Organization (HUGO), it had no real authority, and countries resisted any kind of effort toward enforced cooperation, such as parceling out a chromosome to each country for genomic analysis. Instead, legal warfare over the human genome erupted in 1992 over the issue of whether or not a country could issue patents on fragments of the human genome.

The U.S. government filed patent claims on fragments of about three thousand genes involved in the human brain. European authorities cried foul, as did many American scientists. Although the U.S. government rescinded its patent claims in 1994, it did so in a manner that set the stage for yet further arguments. The U.S. Patent Office had initially rejected the patent claims and was set to reject them on final review, because the claims failed to be specific in terms of exactly which brain function was controlled by which gene. The American scientists had not yet acquired this information—they knew only that certain sequences of DNA (gene fragments) were somehow involved in human brain tissue. While the U.S. government withdrew the patent claims, it did not disavow its reason for seeking to patent the human genome—to ensure that American companies had exclusive rights for a number of years, and a headstart thereafter, in commercially exploiting the human genome. To many, this is simply a form of genomic imperialism.

Companies from the United States, Europe, and Japan are now racing to patent exclusive means of accessing portions of the human genome. They are doing so as part of a quest to control the most important source of wealth and power in the twenty-first century—the Holy Grail of human life. Under the title, "Whose Genes Are They Anyway: Population Groups Can Hold Critical Clues," the highly prestigious journal *Nature,* in the May 1996 issue, described quite candidly how Western companies and governments are scrambling to control the genetic wealth of isolated Third World peoples. Like a replay of the sale of Manhattan to Europeans by the Native Americans for $26 worth of trinkets, Third World locals are offered some agricultural tools or, in one case, $20,000, for their cooperation in providing gene samples that may later be worth hundreds of millions or billions of dollars (rights to an "obesity gene" were recently valued at $70 million). What makes locals' genes so valuable is that they provide a shortcut to identifying the specific gene behind a disease, and hence the winning path to a patent on that gene or on a test or therapy for it. As *Nature* observed:

> [The local] populations often have only a small number of founding members, and a certain degree of inbreeding. Individuals therefore share significant parts of their genomes and genetic alterations predisposing to disease tend to be inherited together with their neighboring DNA. [S]canning the genomes of affected individuals can be an efficient way of mapping genes of interest, sometimes down to a region small enough to allow cloning. . . .

> The advantages of such populations have encouraged scientists and pharmaceutical companies to search systematically for new remote populations with high incidences of particular disorders. One example that promises to help find asthma genes is the highly inbred population of Tristan de Cunha, a small island in the South Atlantic, one third of whose roughly 300 inhabitants suffer from the disease. [C]ountries such as India and China, with remote regions where population groups have little geographical mobility and a high degree of inbreeding, may well prove to be genetic goldmines.

Gold—just what animated the Age of Discovery. Why is there a Human Genome Project? Ultimately, so that one group of people can control the economics of human life: more or less the same kind of reasoning that sent Columbus and others on their missions of imperial conquest. Unfortunately, we all may yet face some of the same deadly consequences felt by other aboriginal peoples, unless a new bioethics of birth comes first.

demographic death: the crime of democide

Demographic death is the extinction of a social group, of a unique piece of human culture. Demographic death is frightening for the same reasons as is the extinction of different species. First, there is a chilling loneliness, a stomach-churning irreversibility, about wiping out something that exists nowhere else in the vast universe. What if we need that species for some presently unknown reason? What if we need that culture for some presently unguessable cause? It's too late once it's gone.

Also, all life is interconnected. Once we extinguish part of life, we set in motion the extinguishment of the rest of life. If we let one species die, it is so much easier to let another species die. If we let too many species die, we destroy the delicate food chain that supports all life on earth. If we let one culture die, it is so much easier to let other cultures die. Once any culture is expendable, all cultures are expendable. All life is interconnected because, otherwise, all life is disrespected. The holocaust of sex described in Chapter 1 is a death of demographics. Hidden within those statistics on the intentional suppression of female births is a bioethical pathogen every bit as vicious as its biological cousins. The bioethical pathogen spreads by justifying subsequent democidal holocausts with those accomplished previously. In other words, if it is all right intentionally to suppress births of XX genomes because of the discrimination women face in life, why not also suppress births of genomes that indicate other conditions frequently discriminated against by society? Solving the problem of discrimination biologically—by eradicating genomes—leads to warfare based on ugly concepts such as "racial hygiene," and "ethnic cleansing." It is far wiser to tackle discrimination with education, to teach that we are not our genome. Any genome can give rise to any kind of person.

The last time European civilization went on an imperial rampage, a great deal of demographic death was caused. Perhaps the best-known example is that of the Native American. An estimated 13,000,000 Native Americans were living at the time Columbus arrived. Fewer than 500,000 live in the Americas today. The general attitude then was that Native Americans were less developed, less perfected humans. To justify their extinction, it was postulated and believed that the world would be no worse off without them. This same kind of thinking animates discussions of genomic imperialism. We may end up being as wrong about genomes as we were about the Toltec, the Navajo, and the Hopi. They knew a lot about avoiding many of the problems that plague urban societies today.

Genomic imperialism is the effort on the part of some genome pools to usurp the space of other genome pools. It is directly analogous to the efforts of Europeans, during the Age of Discovery, to usurp the space of other indigenous peoples around the world. This time around, the imperialists are after biomass, not land mass. As noted by Yale biological anthropologist Jonathan Marks, an offshoot of the Human Genome Project that involves salvaging indigenous peoples' DNA for scanning and decoding could be viewed as: "We've taken your land, we've eradicated your lifeways, we've killed your people, but—guess what—we're going to save your cells."

It is now clear that the stench of demographic death hangs over the genomic battleground—but it can be blown away by the brisk breeze of a new bioethics of birth. Or it can descend like a sulfur cloud and enmesh itself in the very fabric of biotechnology. The choice is ours, as a society, to make over the next very few years. Are we explorers with a human conscience or are we prisoners of our ancient subconscious, of what Carl Sagan calls the "reptilian mind?" If the latter, we are in as much trouble as the dinosaurs, for a reptile with a gene vial is a very dangerous combination indeed.

Most scientists involved with the Human Genome Project feel we are embarking upon a major new evolutionary epoch.

In the words of Robert Sinsheimer, the University of California chancellor who helped start the project, "For the first time in all time, a living creature understands its origin and can undertake to design its future." There are times when technology changes so radically, when it takes such a quantum leap, that we must change some of the basic mores of society in order to keep up. Human genome manipulation is clearly a technological leap, with far-ranging consequences for peace on earth. Accordingly, the lessons of history implore us to adopt new social rules about what one can or cannot do regarding the seeds of sex. In short, it is now time for a fundamentally new bioethics of birth.

It will not be the first time that strikingly new social rules have grown out of major technological advances. The Industrial Revolution, with its endless appetite for both human labor and human consumption, spawned new social obligations to treat people impartially, regardless of their sex, race, religion, or national origin. The genomic revolution, with its endless ability to reshape the human body and mind, must spawn similar new social obligations on the genomic level. People are not their genome, they are independent souls with equivalent rights to respect and dignity regardless of chromosomal configurations.

part II the biotechnology of birth

I have a not so sci-fi fantasy in which a woman ten years hence will tell me that she could not possibly risk having a child by 'in-body fertilization.' . . . How will I explain to this woman why I am troubled by this [in-vitro fertilization], by then routine, way of producing babies? We will live in different worlds; I in one in which I continue to look upon childbearing as a healthy, normal function that can sometimes go wrong but usually doesn't. She will live in a world in which the ability to plan procreation means using all available medical techniques to try to avoid the possibility of biological malfunction. (p.35)

RUTH HUBBARD

The Politics of Women's Biology, 1990

There is no more wondrous biotechnology than that of the human reproductive system. Millions of sperm, each of which holds half a genetic blueprint for a person, are pumped out of the penis in a flow of liquid designed to smooth their passage into another body. Meanwhile, one of a million egg cells, each of which also carries half a genetic blueprint for a person, bursts out of an ovarian follicle and is ushered into the cavernous (from the pinpoint egg's perspective!) Fallopian tube. There, in the moist warmth of our natural petri dish, magical waves somehow decloak the ovum (egg), direct the sperm to it, and, if a merger of sperm and ovum occurs, nurture the zygote (egg–sperm pair) as it glides toward a resting spot on the uterine wall.

As the zygote self-replicates and differentiates (makes cellular copies of itself and orders some of those copies to specialize as particular kinds of cells), we gradually rename the parasite—first we call it a multicelled blastocyst, then a jellybean-sized embryo, and eventually a protohuman fetus. Somehow, we don't know how, different cellular copies activate different biotechnical instructions out of the tens of thousands of instructions contained within each cell's replica of the original genetic blueprint. Once they so specialize, they make copies only of their specialized version of the human cell—even though they keep tucked away within their microscopic pockets the blueprint for the entire human body. After nine months of replication and specialization, there is a neonatal infant comprised of thousands of different-looking cellular carriers of the same genetic blueprint created back at the zygote stage. That genetic blueprint is the baby's human genome. Once the baby is born, the parasitism *really* begins!

Our natural biotechnology is not perfect, not a design shop, and certainly not equipped to produce on demand. For all but a few hours per month, the Fallopian tubes are closed to sperm business. A plug at one end of the tubes blocks their entrance with a thick wall of mucus. But once an ovum enters the Fallopian tubes, the mucus becomes clear as glass and sperm can pass through. Even if sperm and ovum find each other at the right time in the right place, there is a 50 percent chance (some researchers say 85 percent chance) the zygote just won't turn on. The blueprint doesn't make sense, signals get crossed, whatever—nothing comes of the fertilized pair. Not hearing the sound of new life, the body's biochemistry proceeds as usual with its monthly menstrual wash, the zygote being just a speck of flotsam. And as to those relatively rare occasions when everyone gets to the party on time and the party actually starts to happen—about one out of ten of those parties breaks up prematurely. Some specialized instructions don't work, signals get crossed, whatever—the construction site is leveled. Notwithstanding the fully activated menstrual shut-off valve, the embryo still miscarries.

The famous prenatal photographer, Lennart Nilsson, introduces his book, *A Child Is Born,* with the observation that:

> The unborn child is a person no one knows. Will it be a boy or a girl? Will it be dark-haired or blond, tall or short? It has no name and no face. Even the woman who is carrying the child knows only whether it is lively or the quiet, leisurely type. People around her see only a pregnant woman.

Nilsson was right—and wrong. The unborn child is very often known quite well. We often try hard to know the unborn child, through selection of our sperm and our mate, of the child's name, and its DNA. Many people do not see a pregnant woman, but a welfare cheat, baby-boy machine, producer of enemy soldiers, or polluter of the human genome. On the other hand, notwithstanding genomic detective efforts or virulent prejudices, Chinese parents who think they are aborting a baby girl sometimes, to their great distress, abort a baby boy. Blond-haired Germans do give birth to dark-haired kids. Welfare mothers do beget millionaire sons. Geniuses can produce slow learners. A zygote's DNA may be screened for one disease only to manifest a completely different disease soon after birth.

Our natural biotechnology of birth clearly is not a designer's dream. Nor a mass-production shop. Nor perfect. And this is what has given rise to a new scientific biotechnology of birth, based on knowledge gained from the Human Genome Project. From personal eugenics, to social genomics, to the creation of new transgenic peoples: The biotechnology of birth is set to explode with possibilities benevolent, barbarous, and bizarre.

chapter 3 personal eugenics: my perfect baby

Man in the distant future will be far more perfect than he now is.

CHARLES DARWIN
1887

Today it is possible to select, on a limited scale, the demographics of one's offspring. There are a number of ways this can be done. Each method of self-selected demographics—in essence, personal eugenics—needs to be understood in some depth in order to provide a logically consistent and fair bioethics of birth. The possible kinds of self-selected demographics include:

- adoption
- mate selection
- sperm differentiation
- sperm banking
- egg banking
- selective fertilization
- selective abortion
- selective infanticide
- cloning

It is possible, with a lot of perseverence and money, to adopt children from almost any part of the world. In so doing, one is engaging in self-selected demographics. Among the questions would-be adoptive parents will ask themselves, or be asked by others, are, what do we want our baby to look like and what apparent sex do we want? If an Asian-looking child is desired, the parents will probably approach an adoption agency that is active in Asia. If a European-looking child is wanted, the parents may have to look harder and wait longer because those demographics are more in demand. Almost all adoption agencies will ask whether you want a boy or a girl. Clearly, by specifying adoptive child characteristics the parents are engaging in self-selected demographics. It is a biotechnology of birth of the supermarket variety. Instead of actually growing our food, we select food grown elsewhere. Whether parents engineer their own genes to produce a blond-haired, blue-eyed son, or insist on adopting one, a self-selected demographic choice is being made just the same.

In choosing a mate, one can, to a large extent, guesstimate at least the outward appearance of one's children, except for sexual body-type, which remains a 50–50 toss-up. When Asian people mate with Asians, rather than with Africans or Europeans, they are implicitly or explicitly deciding that they want their children to look Asian. Similarly, when two people with different epidermal characteristics mate, they are implicitly or explicitly deciding that they want "mixed" kids. Mating is a biotechnology of the sidewalk café variety. Some persons just fall in love and don't give a second thought to what their children will look like. That's just being so hungry they'll stop at the first café and let the menu surprise them. Others deliberately walk and walk until they find the tandoori or tapas or trout they've set their hearts on. Whether a Kikuyu engineers her own genes to produce a blue-eyed African, or simply marries a Swede, a personal eugenic choice is being made.

It is also possible to defeat the 50–50 toss-up of X or Y chromosomes using any of several current biotechnologies of birth. Some women carefully measure the acid–alkaline balance of their reproductive tract in order, they hope, to kill off the sperm that carry either blue or pink versions of the human genome. These women follow a long and (ig)noble tradition of sperm-differentiation biotechnology. For thousands of years, women in Asia and elsewhere tried to capitalize on nature's slight preference for the survival of Y-loaded sperm. Women have been brainwashed into eating various concoctions, including phallic meats, toxic minerals, and powdered stones chipped off statues of Buddha (presumably why so many of the phallic-shaped noses are missing), to make their bodies produce sons. Doctors prescribe various medicines to increase alkalinity and thus heighten the prospects for Y-carrying sperm. The latest advance has come from Gametrics Ltd., whose biotechnology separates a man's ejaculate into two batches, one with X blueprints and the other with Y blueprints.

Sperm differentiation for sex-determination purposes is surely a form of personal eugenics. A decision has clearly been made that a baby of one sex is not as good, or as desirable, or as perfect as a baby of the other sex—and something genetic is going to be done about it. It has not raised bioethical questions from antiabortion advocates because they do not consider unfertilized sperm to have the same life status as a zygote. Since virtually all of a man's sperm die naturally, consciously killing off (via sperm differentiation) all of the X-carrying sperm does not bother the antiabortion lobby. On the other hand, there are feminists groups that vigorously oppose sperm differentiation. In the words of Vibhuti Patel, whose Women's Center leads the fight against demographic engineering in Bombay, India, "For us, it's the survival of women that's at stake. The social implications of sex-selection are disastrous. It's a further degradation of the status of women."

By combining mate selection with sperm differentiation technology we are moving farther along the slope of personal eugenics. It is now possible to forecast such features as hair color, skin tone, size, and maybe intellectual nature, and to be pretty damn certain about sex. Less and less is being left to nature. The reproductive biotechnology of our bodies is becoming very fine tuned.

Sperm and egg banking offers further refinements of personal eugenics. With sperm and egg banking a customer looks through a menu of demographic possibilities. The table on the facing page displays a typical selection from a sperm menu.

Other sperm or egg banks also offer psychological test scores, detailed heath information, and an option for the donor to have personal contact with the offspring. The customer makes a selection and receives instruction or assistance in fertilizing the banked sperm or egg. Fertility banks offer their services to four categories of customers: persons with conception problems, single adults, gay couples, and heterosexual parents who want different demographics in their child from those they are likely to produce. It is the last category that presents a heightened level of personal eugenics, because an explicit decision is being made to produce a child with different demographics from those that would come "naturally" from one's relationship.

One much publicized example of using sperm and egg banks for personal eugenics was that of a dark-skinned African wife and her light-skinned Italian husband who purchased a Caucasian egg. The egg and the husband's sperm were fertilized in a fertility lab and then implanted in the wife's uterus for normal gestation. When the media asked the couple why they chose a Caucasian egg, their reply was that

Sample Listing from a Sperm Bank Catalog

Race	*Blood Type*	*Ethnic Origin*	*Ht.*	*Wt.*	*Build*	*Skin*	*Eyes*	*Hair*	*Yrs. at Univ.*	*Occup.*	*Hobbies*
Cauc.	O+	Irish–Ger./Eng.	5'10"	130	Light	Fair	Green	Blond/Straight	1	Stud./Bus.	Music/Skiing
Cauc.	O+	Russ. Jew/Ger. Jew	5'9"	165	Med.	Fair	Brown	Black/Wavy	4	Stud./Pol.	Tennis/Fishing
Black	AB+	Afr. Am./Creole	5'11"	185	Med.	Dark	Brown	Black/Curly	4	Civil Eng.	Sports/Poetry
Hispan.	A+	Mex./Puerto Rican	5'4"	158	Med.	Olive	Brown	Black/Straight	1	Stud./Comp.	Cars/Sports
Mixed	B+	Jap./Hispan.	5'9"	165	Med.	Olive	Brown	Black/Wavy	5	Comp. Prog.	Sports/Family
Cauc.	A+	Eng./Ger.	6'3"	160	Med.	Fair	Blue	Blond/Straight	2	Stud./Psych.	Motorcycling
Black	O+	Adopted	6'1"	190	Light	Med.	Hazel	Brown/Curly	8	MD	Travel/Movies

they did not want to subject their child to as much prejudice as she or he would suffer if born with darker skin. Other, more common examples, include eugenics-oriented parents who purchase sperm from donors with high IQ scores.

The use of sperm and egg banks has raised few bioethical problems, except when they are used to contravene widely held social prejudices about interracial or same sex parents. Thus, the concerns have not been eugenic in nature, but more monoracist or heterosexist. By combining sperm or egg banks with sperm-differentiation technology, it is possible to select the sex and likely demographic characteristics of one's offspring regardless of one's own or one's lifemate's characteristics. For example, by purchasing both sperm and egg cells, blending them in a fertility lab and implanting them in oneself or in a surrogate mother, the choice of a child's demographics becomes eugenically arbitrary (separated from the parent's demographics).

A more advanced form of personal eugenics is selective fertilization. With this technique, several zygotes are created from a couple's eggs and sperm in a fertility laboratory. Each such zygote represents a unique shuffling of the DNA code, with half the cards contributed by each mate. All these zygotes are allowed to grow to a several-cell stage (blastocyst), at which time one cell is cleaved off each for chromosomal analysis. The parents are then told about the likely characteristics of each person latent in those chromosomes. Today, they can be told which, if any, will have certain debilitating diseases. Tomorrow, they will also be told which ones are likely to be gay, rebellious, or blond. The parents then tell the fertility lab which human seedlings to dispose of, or to freeze for later consideration, and one of the human seedlings is implanted in a uterus for gestation and birth.

Selective fertilization enables a couple to improve its eugenic success rate significantly over that available merely through mate selection, sperm differentiation, or sperm/egg banking. In terms of the chromosomal make-up of the offspring, the guesstimates of deliberate mate selection are replaced with highly accurate predictions. Basically, selective fertilization turns personal eugenics into a several-hand card game played with all the cards face up.

Selective abortion is similar to selective fertilization except for the manner in which the demographic choice is learned and carried out. With this more primitive approach, a fetus's chromosomes are examined from the surrounding amniotic fluid. Usually all that is looked for is whether it is a girl (XX) or a boy (XY). Chapter 1 showed that in Asia, if it's a girl, the baby is often aborted. Prior to the use of amniocentesis, the same outcome was obtained after birth by selective infanticide.

Finally, the near future includes the prospect of cloning, a technique by which parents can make a child look, and perhaps behave, almost exactly like one of themselves or someone else. When Scotland's Roslin Institute announced, in February 1997, they had created a replica sheep "Dolly" from a single cell of another sheep, all the world knew that personal eugenics had become a new kind of game: Children could become physical copies of the people we liked. It is important to remember, though, that a cloned person is still a unique individual. My IBM PC may look just like yours, but its contents reflect its unique Internet downloads and word-processed documents. In other words, personal eugenics via cloning does not deny individualism.

As wondrous as our natural biotechnology may be, history is replete with human efforts to at least manage, if not control, the biotechnology of reproduction. It is apparently an unshakable part of human nature to want some control over the characteristics of our offspring. From the individual parent's standpoint, personal eugenics must be considered to be a fundamental human right. How else can one explain all the efforts people have gone through to get their children "just right"? These efforts have ranged from specifying the sex and nationality of a child to be adopted to purchasing the sperm and/or eggs of persons of specifically desired demographics for use in creating one's own children. Personal eugenics is as old as royal families breeding only with royal blood and as timeworn as replacing female infanticide with X-versus-Y sperm differentiation to achieve male heirs. It would be as impossible to prevent the practice of personal eugenics as it would be to prevent people from choosing their own mates. The pressing question in the age of genomics is whether there should be any limits to personal eugenics. Is even cloning allowed?

There are, today, no internationally enforceable legal limits on personal eugenics. Genetically blind or deaf individuals, for example, are fully within their rights to produce genetically blind or deaf children. Limitations on the use of sperm and egg banks and artificial fertility techniques do exist in some jurisdictions, but these limits or bans can be avoided simply by going elsewhere for conception technology. The current state of the law can be summarized by saying that anything people do naturally to achieve eugenic (or diseugenic) objectives is legal, and anything people do with artificial biotechnology to achieve eugenic outcomes is also legal so long as no biological harm is caused to the newborn. The United Nations, for example, has proposed in its Declaration on the Human Genome and Human Rights to prohibit only "any eugenic practice that runs counter to human dignity and human rights." The UN's International Bioethics

Committee explained that "non-therapeutic interventions" (e.g., sex selection for families with three children of the same sex) do "not appear to be a clear breach of respect for human dignity." Given the vagueness of "human dignity," it is clear that the UN's International Bioethics Committee—headed by French Constitutional Court member Noëlle Lenoir and staffed by globally representative bioethicists and Nobel Laureates—has endorsed personal eugenics.

The debate over personal eugenics will revolve around the meaning of the terms "biological harm" and "human dignity." For example, is it harmful to manipulate a zygote's genes so that a newborn has a third eye in the back of its head? Obviously, the consensus of opinion today is that such a manipulation by creating a "freak" would be harmful. In an objective sense, however, the person might not be worse off, and certainly could offer an athletic team a special advantage. Is it harmful to manipulate a set of genes so that a newborn has vastly better eyesight than 20/20, or has superhuman strength, or has an inability to wrinkle, or lose hair, or get fat? Clearly, it is more difficult with these modifications to claim that harm is being caused. And, so long as harm is not being caused, and human dignity is not sacrificed, international law does not prevent this kind of personal eugenic engineering from occurring.

Although some have called for a global moratorium on any genetic manipulation that causes inheritable changes, such an approach would seem too broad and contrary to existing human rights. For example, suppose there is a genetic manipulation that results in an inheritable resistance to cancer. Why should not parents be able to pass this trait on to their offspring? International treaties recognize the fundamental right of all peoples to advance the state of human health.

Nearly all of us are personal eugenicists. My idea of perfection is not yours, but virtually all parents look at their newborn baby and feel it is perfect for them. Nearly all of us try, at least a little bit, to get the baby we want. We pick a tall or a cute or a happy mate and hope our children will be a bit like him or her. We ask to adopt a blond or a brown or a boy because that's the perfect one for us. Few loving parents would say no to a magic wand that would make sure their offspring didn't suffer from any number of diseases as they grew up. Selective fertilization can be that magic wand. Few loving parents would say no to a magic potion that would ensure their son never balds, their daughter never wrinkles, and all their kids have perfect pitch (musical or baseball). Genetic engineering may be that magic potion.

The magic wands and magic potions are not yet fully ready, but they soon will be. Let's not throw them all away for fear of what we are doing to the human race. The human race has always changed genetically, and it is changing more rapidly now than ever before. We should throw away any magic wands or potions that cause pain and suffering, for to permit pain to occur unnecessarily is cruelty. Still, we must be careful to distinguish pain from prejudice. The blind, the deaf, the fat, and the freaks suffer not from pain but from prejudice. Their disease is only society's dis-ease with them.

Personal eugenics is the right to birth the children that you choose. Personal eugenics is not the right to impose your view of perfection on others. That brings us to the realm of social eugenics and demographic engineering. While we have reached agreement that society should limit personal eugenic choices to those that do not cause intrinsic harm, others foresee a much broader mandate for social control. It is to this realm of social eugenics that we now turn.

chapter 4 social eugenics: my perfect society

Germans have an abiding and understandable fear of anything to do with genetic research. It is the one science that reminds them all of everything they want to forget.

London Financial Times, May 10, 1989

Those who cannot remember the past are condemned to repeat it.

GEORGE SANTAYANA
The Life of Reason, 1905

We saw in the last chapter that personal eugenics, the right to influence the characteristics of our offspring, is as old as history and as new as the latest biotechnology. The new bioethics of birth guarantees the right of parents to influence the characteristics of their offspring, including the use of biotechnology to improve the children's biosystem. Personal eugenics is a human right so long as it is carried out without destroying intentionally created life and so long as the parents' eugenic vision does not prevent the new person from intrinsically enjoying life.

Social eugenics is the repression of personal eugenics. Social eugenics means that there are rules that affect the ability of parents to produce the children they wish to produce. Consequently, social eugenics is, on its face, contrary to the bioethics of birth. Any ideology contrary to the bioethics of birth facilitates some kind of demographic death. Indeed, every social eugenics movement of the twentieth century has resulted in just such a parade of horrible destruction.

Social eugenics can be traced back at least to ancient Sparta, when, by law or custom, babies who failed to meet certain ideals of health were killed upon birth by exposure. It is doubtful whether a mother who thought her baby was healthy enough would have been allowed to save it from its social eugenic fate—left on a rocky hillside for a vulture to devour. After the Greeks came centuries of socially enforced mating practices, which were crude efforts at eugenics. Even today, opinion polls report that 20 percent of Americans believe it should be illegal for people from different "races" to marry. While such marriages actually were illegal, from around 1900 to 1967 in most states, the United States was practicing a crude form of social eugenics—attempting to suppress the birth of "interracial" children.

The modern heritage of demographic engineering is usually traced to Sir Francis Galton, who coined the term "eugenics" in the late nineteenth century. Galton, a nephew of Charles Darwin, considered eugenics to be a straightforward application to humanity of the natural selection principles his uncle had enunciated for the plant and animal worlds. Galton bootstrapped himself up Darwin's mantle of fame by suggesting that humans, unlike his uncle's subjects, could intentionally control the process of natural selection. For example, if only the tall mated and had children, Galton reasoned, then the human race would grow taller.

Unfortunately, most of the followers of Galton, up to and including Richard J. Herrnstein and Charles Murray, the authors of a much-publicized and frankly pseudoscientific tract called *The Bell Curve* (1994), make a major human classification error that Darwin himself knew enough to avoid. That error is to subdivide humanity into "races" based on skin tone, hair texture, and facial features, and to assume that certain behavioral genes go with certain "races." Scientists are clearer now than ever that "race" as society knows it—black, white, and so on—is a political creation, not a genetic reality. We are fooled by our eyes that the dozen or so genes that determine melanin content or hair texture can be lumped into five or so racial categories. Racists then exploit our reliance on visual misperception to argue that each racial category pulls behind itself hundreds of behavioral or intelligence genes. But Luigi Luca Cavalli-Sforza, the world's leading gene demographer, after comparing the genes of thousands of living and deceased people from every part of the world, has clearly documented in his monumental 1994 treatise, *The History and Geography of Human Genes,* that it's just not so:

> The classification into races has proved to be a futile exercise for reasons that were already clear to Darwin [who observed that all human peoples "graduate into each other"]. . . . From a scientific point of view, the concept of race has failed to obtain any consensus; none is likely, given the gradual variation in existence. It may be objected that the racial stereotypes have a consistency that allows even the layman to classify individuals. However, the major stereotypes, all based on skin color, hair color and form, and facial traits, reflect superficial differences that are not confirmed by deeper analysis with more reli-

> able genetic traits. . . . [W]e can identify "clusters" of populations and order them in a hierarchy that we believe represents the history of fissions in the expansion to the whole world of anatomically modern humans. At no level can clusters be identified with races, since every level of clustering would determine a different partition and there is no biological reason to prefer a particular one. . . . There is no scientific basis to the belief of genetically determined superiority of one population over another.

In a nutshell, Professor Cavalli-Sforza is observing that skin tone is just one of many genetic traits that are shared by groups of people worldwide. Other traits, usually biochemical ones, like blood type, that we can't easily see, provide a more consistent means of grouping people than does skin tone. And no genetic grouping factor provides any community with a basis for claiming superiority. Certain individuals may have great attributes, but those attributes can be found randomly in every grouping of people.

Notwithstanding the lack of any consistent scientific basis, Galton's writings spurred eugenics theorists worldwide to develop various approaches to "improving" the human race. During the twentieth century, these approaches have fallen into three areas: positive eugenics, negative eugenics and medical eugenics.

positive eugenics: grow the genome

Positive eugenics is often presented as the friendly face of demographic engineering. The key to this approach is to give incentives to persons with socially popular characteristics to produce more children. Positive eugenics has taken form as:

1920s era "fitter family" contests in Middle America, in which prizes were given to families (always Euro-American) that were large, mostly male, and physically "normal";

Nazi-era cash awards and vacations to Aryan women who bore Aryan-looking children, with the awards growing larger as ever more Aryans were produced;

Singapore's 1994 law to reward college graduates who produce children with a greater array of social benefits than nongraduates who produce children.

George Bernard Shaw, a rabid positive eugenicist, quipped that "we have never deliberately called a human being into existence for the sake of civilization, but we have wiped out millions." With perhaps greater accuracy, he also commented that the British Empire went to great lengths to exterminate indigenous peoples but did very little to ensure that the English colonists were worthy of their conquered lands.

In fact, positive eugenicists have always run into serious roadblocks in implementing their demographic dreams. First, no amount of incentives ever seems to be enough to persuade adequate numbers of "genetically desirable" females to spend a considerable portion of their lives breeding and nursing infants. They are simply too sensible. To the contrary, statistics repeatedly show that as a woman's education and economic standing go up, her birthrate goes down. As indicated in the World Bank's "Investing in Health" annual report for 1995, "increasing female literacy rates by 10% is likely to lower the infant mortality rate by an estimated 10%," and "the mortality decline that has occurred almost everywhere has usually been accompanied by steep falls in fertility." Indeed, the World Bank's figures show that in countries where women have largely achieved at least secondary education, birthrates tend to hover around 2 to 3 live births per 100 people per year. Where women are generally held back from education, the annual birthrate ranges from 4 to 7 live births per 100 people. In other words, educated women—those the positive eugenicists are looking for—are unlikely to be willing baby machines.

Second, eugenicists discovered, to their great dismay, the incontrovertible reality of "genetic drift." This phenomenon is the natural tendency of chromosomes to break down the results of selective breeding within a few generations. This is easy to see if you imagine the genetic code to be like a deck of cards. Suppose that some society's vision of a eugenic ideal is a straight flush. With considerable effort, the deck can be arranged so that all the cards are grouped into numerical sequences of the same suit. Now each mating of this eugenic society is like a shuffling of the deck. It is easy to see that with each shuffling the deck gets closer and closer to its original random order. It is much the same with genetics. As a result, positive eugenicists can never rest. Even traits that they mate away will creep back into the gene pool.

Finally, the positive geneticists run into the moral quagmire of family law and religious morals. The logic of positive eugenics would require a genetically desirable male to be able to impregnate as many genetically desirable females as possible. Because males have an essentially unlimited ability to produce sperm, but females can gestate but one child at a time, the logic of positive eugenics would have women in a permanent state of pregnancy. But as we noted above, a "genetically desirable" female would never agree to such a life. Also, the positive eugenics ideal is not being achieved if reproduction is limited to a single female's birthrate. There are only two positive eugenics alternatives to this dilemma. One is that a class of nongenetically desirable females can serve as mothers for the class of genetically desirable sperm donors. This creates a family law and morals problem (polygamy) and also dilutes the eugenic contribution of the male with a normal contribution from the female. A second solution is for a class of surrogate mothers to serve as gestating machines for genetically desirable females fertilized by genetically desirable males. This creates family law problems of parental rights (surrogate versus egg–sperm donor) and polyandry (if multiple men are involved), not to mention serious theological objections. In any event, there are very few surrogate mothers available.

Despite the logical problems inherent in positive eugenics, its proponents have taken it as far as the establishment of a sperm bank for geniuses. The hope here is that at least some women will pay $3,000 ("regular" sperm costs about $250) to be inseminated with "genetically desirable" sperm, and in that way the genetic stock of the human race will be at least partially improved. In the words of a spokesperson for this sperm bank, the Repository for Germinal Choice: "We enable outstanding men to have more offspring than they would have otherwise. This puts more genes from some of our best men into the human gene pool."

With only a few hundred intentionally bred offspring after three decades in operation, it seems unlikely that the positive eugenicists' dream will get too far. Frustrated by uncooperative females, unruly chromosomes, and unfriendly laws, the positive eugenicists frequently mutate into something more sinister. Not patient enough for millennia of incentivized demographics, many social eugenicists decide to engineer humanity by what might be called, without mincing words, slash-and-burn techniques.

negative eugenics: gas the genome

A worst-case scenario for negative eugenics can be found in the James Bond film *Moonraker*. Disturbed by his inability to control the complex world, a megalomaniac billionaire, Mr. Drax, hatches a plot to wipe out human (but not plant or animal) life on earth using deadly microbes. Meanwhile, he sequesters a small coterie of eugenically selected men and women on board a space station he constructs, safely out of the mass extermination on earth. Before he can repopulate the world with his eugenically selected clan, however, Agent 007 blows up the space station and saves the planet. Drax's original plot to wipe out "inferior life" is pure negative eugenics.

Space-station holding areas are not generally available, although another real-life billionaire, Ted Bass, has created a similar, environmentally sealed structure in the Arizona desert. Known as "Biosphere II," it is advertised as being able to support people sealed off from the environment for one to two years, and is actually being offered for sale. Biospheres notwithstanding, even the most genocidal negative eugenicists must exercise considerable caution lest they wipe themselves out. History's most famous negative eugenicist, Adolf Hitler, failed to extinguish all the genes carried by Jews, Gypsies, and the differently abled—but he did wipe out millions of people, including Aryans and himself, in the process.

Other negative eugenicists have been considerably more cautious, usually wiping out only the unborn descendants of genetic "undesirables," not actual living, breathing people. Sterilization has been the favored tool of negative eugenicists. The United States and Germany are the two countries with the most active records of the use of sterilization for eugenic purposes.

Eugenic sterilization first gained popularity in the United States shortly after the First World War. The public was led to believe that social undesirability, often called "feeblemindedness" at that time, was an inherited trait. Furthermore, negative eugenics advocates propagated figures showing that, while it would cost only about $100 to sterilize an undesirable, it would cost millions of dollars to deal with all of the costs of their offspring (crime, welfare, incarceration, hospitalization). Based on this logic, over half the American states adopted eugenic sterilization laws. Among the thousands of persons sterilized was a wide array of nonconformists, including sexually active young women, petty criminals, and those unable to score well on IQ tests.

There were some civil libertarians who opposed the eugenics laws. These individuals doubted that "feeblemindedness" was necessarily an inherited trait, worried about ever-widening boundaries for who might be sterilized, and questioned whether it was environment, not genetics, that affected the quality of social life. These voices of doubt were given short shrift in the 1927 Supreme Court case that determined the constitutionality of negative eugenics, *Buck v. Bell.*

Carrie Buck was an eighteen-year-old institutionalized woman when she became pregnant. She and her mother, who was also institutionalized, both scored very low on IQ tests, and Warden Bell was not happy to see her pregnant. The decision to apply a recent eugenics law to her, and thereby stop any further pregnancies, was appealed to the United States Supreme Court. Carrie Buck's attorneys relayed the civil libertarians' warning that if the sterilization was upheld, "a reign of doctors will be inaugurated and in the name of science new classes will be added, even races may be brought within the scope of such a regulation and the worst forms of tyranny practiced."

Carrie Buck had given birth to a daughter, Vivian, who was placed in foster care. When Vivian was only seven months old, a social worker testified that "she didn't look quite normal." The government's leading eugenics expert swore that the Bucks "belong to the shiftless, ignorant, and worthless class of anti-social whites of the South." By a vote of 8-to-1, the Court upheld the constitutionality of negative eugenics generally, and of its particular application to Ms. Buck. In the words of no less than the great jurist Oliver Wendell Holmes:

> We have seen more than once that the public welfare may call upon the best citizens for their lives. It would be strange indeed if it could not call upon those who sap the strength of the State for these lesser sacrifices in order to prevent our being swamped with incompetence. The principle that sustains compulsory vaccination is broad enough to cover cutting the Fallopian tubes. Three generations of imbeciles are enough.

The decision in *Buck v. Bell* provided renewed impetus to negative eugenics movements during the 1930s in the United States and Germany. Indeed, the leading eugenics expert, who had provided the *Buck v. Bell* testimony regarding the "worthless class of anti-social whites of the South," received an honorary degree from Hitler's Germany, which he accepted as "evidence of a common understanding of German and American scientists of the nature of eugenics." And, as the Supreme Court had been warned, the scope of sterilization spread quickly. Indeed, genetics historian Daniel Kevles observes in his 1986 book, *In the Name of Eugenics,* that German eugenicists said "they owed a great debt to American precedent."

In the United States, up to half the sterilized persons in some states were African American, while in Germany forced sterilization had a quarter-million victims within three years after passage of the Eugenic Sterilization Act of 1933. Germany's law initially covered feeblemindedness, mental disease, blindness, and various addictions, deformities, and other hereditary conditions. Sterilization of diabetics, Jews, and "mixed-race" (African–German) children was soon added in the interest of "racial health." At the Nuremberg trials after World War II, one Auschwitz survivor testified that women were used as guinea pigs for testing rapid sterilization by Xrays, injections, and electrical shocks so that the Germans could find a rapid enough method "to repopulate all western European countries with Germans within one generation after the war." Apparently, Mr. Drax was not the first to try taking negative eugenics to its logical limit.

Although some argue that negative eugenics need not lead to holocausts, the historical and logical fact is that it does. After sterilizing *some* of those with purportedly hereditary weaknesses, logic asks why not sterilize *all* of those with purportedly hereditary weaknesses? If one believes that there are significant genetic differences between social groups of people (Asians, Africans, Aryans), as do contemporaries such as the authors of *The Bell Curve,* logic next extends sterilization to those social groups of people with purported hereditary weaknesses. For example, the *Bell Curve* authors argue that African and Hispanic Americans on average have less genetic intelligence than do Euro-Americans. A negative eugenicist would logically sterilize the *Bell Curve*'s rejects along with diabetics, epileptics, and alcoholics.

Six years after Germany passed its Eugenic Sterlization Act, euthanasia was inaugurated on the same classes of people earlier scheduled for sterilization. The logic of negative eugenics inevitably leads to, "If we don't want your offspring, we don't want you." The Nazis medicalized their bigotry with a concept of "lives not worth living," based on the title of a guiding 1920 manuscript, "The Release for Destruction of Lives Not Worth Living." The short hop from euthanasia to genocide is clearly explained by Harvard Professor Emerita Ruth Hubbard in her book, *The Politics of Women's Biology:*

> By September 1941 over seventy thousand ["lives not worth living"] inmates had been killed at some of the most distinguished psychiatric hospitals in Germany, which had been equipped for this purpose with gas chambers, disguised as showers, and with crematoria. (When the mass extermination of Jews and other "undesirables" began shortly thereafter, these gas chambers were shipped east and installed at Auschwitz and other extermination camps.) Most patients were gassed or killed by injection with lethal drugs, but a few physicians were reluctant to intervene so actively and let children die of slow starvation and the infectious diseases to which they became susceptible, referring to this as death from "natural" causes. . . . There is a direct link between this campaign of "selection and eradication" and the subsequent genocide of Jews, gypsies, communists, homosexuals, and other "undesirables." Early on these people were described as "diseased" and their presence as an in-

> fection or a cancer in the body of the Volk. . . . This rationalization [is] "the medicalization of antisemitism."

Hubbard points out clearly that the bigotry of the Nazis was carefully justified with medicogenetic "evidence." The Nazis claimed only to be carrying out what was already established as "medical truth." This is why we must exercise the utmost vigilance in refusing to give any credence to purportedly scientific evidence that some genomes are better than other genomes. To do so leads to efforts to suppress the reproduction of the allegedly "worse" genomes. And, quoting again from Daniel Kevles, "A river of blood would eventually run from the sterilization law of 1933 to Auschwitz and Buchenwald."

For several decades after World War II, negative eugenics fell out of favor. People didn't want to be associated with Nazism, and the public grew to understand that "feeblemindedness" was too complex a set of traits to be "inherited." For example, Carrie Buck's daughter turned out to be, in her teacher's word, "bright." With Hitler's fade into history, however, and the rise of the Human Genome Project, negative eugenics is once again gaining ground. As we move into the twenty-first century, slash-and-burn demographic engineering is a major threat to personal eugenic freedom and to human diversity.

The Human Genome Project has already enabled us to identify dozens of medical and demographic conditions based on analyzing the chromosomes of a newly formed embryo or a gestating fetus. It is now possible for parents to say that they don't like the hand they were dealt, and so they will abort and start over again. A great many Down's syndrome fetuses are aborted in this manner; fetuses with Tay-Sachs disease are also aborted. Millions of people would abort bearers of gay or obese or manic depressive genes if they could. If mandated by society, then this, too, is negative eugenics—achieving a "more perfect" populace by suppressing the births that society doesn't want.

With the human genome fully mapped, all human characteristics are identifiable via genetic testing. It is reasonable to expect that an embryo could be genetically tested for any of dozens of different diseases, predispositions, and phenotypic (body-type) characteristics. Indeed, such embryo-testing services were first offered for a few select genes in the mid-1990s under the rubric "pre-implantation diagnosis" (PID). In vitro fertilized embryos with as few as 8 cells are analyzed and, if the genetic configuration preordains a disease, discarded. A new technique called "uterine lavage" enables the embryo to be washed out of the uterus, tested for various gene sequences, and reimplanted during the next menstrual cycle. Alternatively, the amniotic fluid could be sampled for genetic testing during the first trimester of pregnancy, or perhaps wayward fetal blood cells could be culled from a maternal blood test. The point is that it will be easy to scrutinize the seeds of sex.

A government that wanted to reduce its health care costs—now the largest segment of many governments' budgets—could pass a law mandating the suppression of births for any of a number of eugenic reasons. Some of those reasons may appear to be medical, such as reducing the number of carriers of sickle cell genes—but would actually be democidal—reducing the numbers of African Americans. Such a program is not currently illegal under international law.

A government that wanted to homogenize its "family values" could pass a law mandating the suppression of births of persons likely to express homosexual, transsexual, or bisexual orientations. This might occur if it were found that some set of genetic markers correlated highly with sexual orientation. A legislative body might be persuaded that it was better to "nip the gay problem in the bud" than to deal with same-sex marriage laws and nonheterosexual role models for youth. Such a governement genetic-testing program could be couched in terms of preventive medicine, because doctors in some countries diagnose diverse sexual orientations as mental illness. The effect of such a government genetic-testing program could be to wipe out an entire demographic group—gays, lesbians, and transsexuals—in a single generation.

The preceding examples are intended to show that negative eugenics is a current threat: Hitler lives. Although we all hope that the Auschwitz gas chambers are forever behind us, much the same outcome can be produced through genetic testing and selective fertilization or abortion. A global Genocide Treaty outlaws mass murder, and yet the holocausts of Cambodia and Rwanda still occurred. But not even the Genocide Treaty bans negative eugenics for medical or nonracial reasons. Then what is to stop, or even slow down, holocausts of particular genomic characteristics once the Human Genome Project is complete? How do we ensure that the fields of biotechnology make room for all the seedlings of sex, not only for a few select strains?

medical eugenics: cure the genome

In the words of bioethicist Arthur Caplan, American eugenics is no longer "going to come from a Hitlerian dictator saying, 'You must do this.' It's probably going to come from a society saying, 'You can have a kid like that if you want, but I'm not paying.'" For example, a health insurance program might oblige its users to submit to genetic testing and birth counseling in exchange for health insurance, or give them incentives to do so with lower rates. After all, the argument will go, medical resources are limited; it is not fair for one couple to use a disproportionate amount of these resources on a very sick baby whose birth could have been averted for the good of the state.

Health insurance programs might even require citizens at a heightened risk of genetic disease to use artificial fertilization techniques, so that embryos that evidence disease-bearing chromosomes could be aborted in the petri dish. In such practices, negative eugenics would clearly be at play, with certain chromosome configurations being suppressed.

It is more difficult to argue against selective fertilization when a clear-cut, life-threatening disease is involved. Our sympathies are probably not with some social eugenics goal, only with avoiding pain and anguish for the newborn and its parents. The government or health care program may merely make it easier for us to do what we would want to do anyway as a matter of personal eugenics. But what happens when the list of conditions we must abort includes obesity, aggressiveness, or homosexuality? Are we still doing the right thing for the child, or are we participating in a misguided attempt to engineer human demography? Are we simply practicing predictive medicine, or are we inaugurating a new kind of tyranny of racial hygiene?

Certain researchers argue that, on average, some races are less "intelligent," and therefore less able to "thrive," than others. Could a government mandate, for "medical reasons," the selective abortion of any embryo lacking certain chromosomal "intelligence" markers? If not for the first child, what about for the second or third child? Regardless of whether or not the so-called chromosomal intelligence markers turn out to be accurate or meaningful, could this not be a back-door method for suppressing one or another racial or ethnic group? If the coded chromosome is not intelligence, what if it is some other condition that is more prevalent in one racial or ethnic group than another?

The suppression of intentionally conceived genomes, for any reason, is fraught with ethical concerns. There are gray lines everywhere. Between personal and socially mandated eugenic decisions. Between medical and racial boundaries. Whether we promote a birth or suppress a birth, we are making a eugenic choice. Each individual has a right to make that choice for his or her own offspring. Governments may have a right to make choices that clearly affect their expenditures. But negative social eugenicists should proceed with great caution. In the words of historian of science Ruth Schwartz Cowan, "If nothing else, the history of the twentieth century ought to have taught us that individuals can sometimes behave badly, but they can never behave as badly, or as destructively, as governments can."

In modern society, medicine is an arm of the government. Whether or not public health insurance exists formally, the medical and insurance professions are comprehensively under the regulatory power of the government. Accordingly, if the insurance industry "behaves badly," it is tantamount to the government's "behaving badly." Demographic destruction by medical proxy is demographic destruction just the same.

As a step toward this kind of democide, Dr. Jacques Testart, director of gamete maturation and fertilization research at the French Institut National de la Santé et de la Recherche Médicale, warned in a late 1995 article in the pharmaceutical industry's publication *Genethics* that medical eugenics risks bringing a problematic "health hierarchy" to society:

> A health hierarchy will be set up between, say one individual embryo, fetus or person at 78% risk of heart disease and 59% risk of asthma versus another individual with risks of only 8% and 13%, respectively, thereby inaugurating a revolution in ethics. To date it has been impossible to grade and quantify differences in genetic inheritance. Now, however, we have entered an age in which a prior pre-disease grading in terms of statistical risk can stratify the population along health lines, with potential impact on their status and prerogatives in such areas as education, employment, insurance, procreation, etc. Incorporation of embryos into this creeping health hierarchy will convert the egg, as some doctors have already proclaimed, into "the smallest patient," i.e., an object of medical attention before any intimation of disease. "Treatment" in this case will consist in eliminating the great majority of eggs.

We can no more permit negative eugenics to be practiced by the medical and insurance industries than by tyrants and governments. The purpose of medicine is at least to "do no harm." This injunction must apply to all intentionally conceived life and it must apply to life in the aggregate—demographics—as well as life one by one. The purpose of insurance is to spread the burdens of a few across the shoulders of the many. This mandate must apply to all genomic burdens, and it must apply to the part of the human genome we all share as well as the part of the human genome we each have individually.

chapter 5 transgenic creationism: my perfect monster

I have asked the Miltonic questions Shelley poses in the epigraph of Frankenstein: "Did I request thee, Maker, from my clay to mould me man? Did I solicit thee from darkness to promote me?" With one voice, her monster and I answer "no" without debasing ourselves, for we have done the hard work of constituting ourselves on our own terms, against the natural order. Though we forego the privilege of naturalness, we are not deterred, for we ally ourselves instead with the chaos and blackness from which Nature itself spills forth.

DR. SUSAN STRYKER
queer theory academician, in "My Words to Victor Frankenstein Above the Village of Chamounix," *1994*

We have seen that social eugenics must be stopped lest it unleash the horrors of demographic death upon us all. We have also seen that social eugenics is simply the antithesis of personal eugenics—it is the imposition by society of rules that limit personal eugenic choices. One final question remains:

Are there no limits to the practice of personal eugenics? Are we free to produce any kind of children we want?

Today, by law, no one can tell you to abort a baby because of its genes. While the barking dogs of negative social eugenics are, once again, on the horizon, they are not yet in our midst and may yet be kept at bay. Parents who have been advised that they have a 100 percent chance of producing a child who will be congenitally ill, in some cases, horribly so, cannot be prevented from conceiving that child. Parents could have several reasons for doing so. They may not believe the medical specialists. After all, there are medically diagnosed invalids who learn to walk, untreatable cancer patients who somehow remit, and infertile women who miraculously conceive. The parents may believe the medical specialists, but have a still greater faith in a deity. God may want the child to live to teach us something or may want the child to enter heaven. There is an East African saying, "One must live in order to die."

While the angry negative side of eugenics does not yet obscure our reproductive choices, there are concerns on the positive side. Proposals have been made for a "moratorium on germ line therapies." This means that experts are asking for a freeze on altering the half-blueprints contained within egg and sperm cells ("germ lines"). For example, it may be possible through Human Genome Project research to microsurgically alter the characteristics of an egg cell's DNA to change the likely color, size, health condition, or other characteristics of a person before birth. A moratorium is being requested because scientists do not know the consequences of fiddling with the DNA of a half-cell human. They are fearful that the Human Genome Project will get quashed if germ line therapies are undertaken by a few radicals with regrettable results. And they believe there are safer alternatives.

For example, recently an adrenal hormone has been associated with longevity. While everyone's hormone levels decline with age, people are born with differing hormone-production capabilities. People born producing relatively high amounts will always produce relatively higher amounts throughout their lives. Now, suppose that it turns out that persons' lifespans are determined by how much of this adrenal hormone they produce. A genetics lab might promise persons producing low amounts to actually change their germ cell DNA in such a way as to ensure that their offspring produce high amounts, and thus are likely to live longer. The genetics doctors may plan simply to swap a couple of organic molecules at some precise point on one of the 23 chromosomes. Inadvertently, something slips—as often happens in surgery—and the children are born instead with a excruciatingly painful nervous disorder or some other malady. Not only would everyone in sight be sued, but the entire field of genetics would suffer a public relations earthquake.

On the other hand, a genetics lab might promise a person with low adrenal-hormone production that it will fertilize several germ cells, scan the DNA of each, and implant a fertilized germ cell only if it has high hormone production capability. In this case, it is much less likely that something would go wrong—the excruciating painful nervous disorder—because the geneticist is not actually changing Mother Nature's machinery. The geneticist is simply examining several of Mother Nature's machines, and taking home the one that works best. Nevertheless, the person's germ line is still changed because the offspring's propensity for low hormone production was presumably left with the discarded embryos. The offspring will probably pass on only high hormone levels to their future generations. No lawsuits are likely in this scenario, and hence it is much preferred by geneticists.

Despite the geneticists' preference for implementing personal eugenics choices by growing several embryos and selecting "the best," there is likely to be continued pressure for actually changing the germ line. The reason for this pressure is that most of the time none of several test-tube embryos are going to have the specialized features that parents want for their children—immunity from most diseases, long life, creativity, intelligence, strong physique. One way to accomplish these "designer babies" is with the technology of transgenics, merging portions of genes from different persons or different species. With this technology, children literally have more than two immediate parents—immediate sources of genetic material—and one or more of these extra parents might not even be human.

how transgenics works

Transgenic technology was invented in the 1980s to create biochemically modified forms of existing animals. The goal was to create new forms of animals that not only had one or a few biochemical characteristics of a reproductively incompatible species, but were able to pass these "transgenic" characteristics on to their offspring. For example, mice were frequently used as the basic animal, and various human biochemical characteristics—blood components, particular hormones or enzymes—were cleaved into the mice. The result was a new kind of mouse that bore mice offspring with some specifically human biochemical characteristics. One such mouse was actually deemed worthy of a patent, and it is known today as the Harvard mouse.

The reasons for creating transgenic species are ostensibly always to help people. Perhaps the most common reason is to develop "animal models" of human diseases. This means to create an animal that biochemically responds as a human would to a particular disease, such as cancer or AIDS. Once this is accomplished, it is possible to test various treatments on the animal that would be too risky to test on humans. If the treatments were tested on an animal that was not to some extent humanly transgenic, the test results would be less useful, and often not useful at all. For example, monkeys get the AIDS virus (HIV) but do not get sick from it because their biochemistry is different from that of humans. A humanly transgenic monkey could be made to get as sick from HIV as humans do.

Another reason scientists create transgenic species is to use the offspring as living factories for the production of biochemical substances that humans need. For example, one company, Genzyme Transgenics of Massachusetts, now raises transgenic goats, which secrete rare life-saving human proteins through their milk. Millions of people lack these proteins from birth or due to disease. Another company, DNX Corporation of North Carolina, is raising herds of thousands of transgenic pigs whose veins course with human hemoglobin. These pigs are expected to become part of the nation's emergency blood supply and a replacement for expensive and problematic artificial blood. DNX also operates the National Transgenic Development Facility that creates transgenic animals for customers on a contract basis.

The basic mechanics of transgenics are simple, although the theory is elusive. We know that following certain steps "works," but we don't really know how or why. We are like the ancients who knew that coitus often led to childbirth, but had no idea of the how or why.

Transgenics begins with an idea. In the words of Dr. Frankenstein, what kind of creature do we want to create? Whatever those desired characteristics are—greater milk production, human blood, disease immunity—we must first find the blueprint for those characteristics in something's DNA. The main purpose of the Human Genome Project is to decipher the chromosomal blueprint for everything that is biologically human. Similar genome projects are underway for other animals. As a result, scientists aleady know which parts of which chromosomes are responsible for a wide variety of biochemical capabilities. For example, we already know quite well which part of the human genome is responsible for instructing the body to make hemoglobin.

The next step in transgenics is to separate the part of the DNA blueprint we want from an entire genome, and to make lots of copies of this separated fragment. This technology, called "recombinant DNA," became well understood during the 1980s. Special biochemical "scissors" known as "riflips" were found that could cut the DNA at fairly specific locations. Thus, scientists simply purchase the riflips they need for the part of the chromosome that they want to cut. The riflips work more or less like the "find" or "search" command in a word-processor program. With blinding speed, the riflips flash through the lengthy genetic document and stop at just the right place. Making lots of copies of the desired genetic fragment is also straightforward. The fragment is simply inserted inside a bacterium that duplicates itself, and whatever is inside it, every few seconds. If a cell doubles every minute, in an hour you'll have a zillion of them (1 followed by 18 zeros).

The third step in what can now be called "transgenic creationism" is to harvest and fertilize some eggs and sperm from the species you want to modify. This is standard in-vitro-fertilization stuff for humans and even more routine in the animal husbandry field. Bull sperm was a hot commodity long before the human sperm banks opened for business. The harvested sperm and eggs will find each other in the test tube, and a zygote will result. So step three is simply creating the same two-headed monster in a test tube that nature always creates in utero—a zygote with two nuclei, one holding the egg's half-blueprint and one bearing the sperm's half-blueprint.

On day four, science creates life in its mental image—a syringe full of copies of the desired genetic trait is actually microinjected into the sperm's nucleus. "Ouch" says the sperm, but here is where the magic begins. Somehow, some of the time, one and only one of the injected genetic fragments finds its logical place along the interminable genetic code of the proper chromosome and actually replaces the rightful occupant of the slot. How this "search, find, and replace" process works, we do not know. Why it works sometimes and not at other times, we do not know.

In the fifth step of creation, the egg and sperm merge their genetic codes, and form a one-headed cell with a full, albeit cut-and-pasted, genetic blueprint. From that point on, the crucial question is whether the embryo will turn on—live—or just turn out to be a dud. Even without hypodermic interference, about half of all zygotes don't work. The microinjected gene fragment will find it is dancing with a biochemical stranger from the egg's strand of DNA. A fight big enough to doom the whole zygote might break out. Or they may try to get along. Or one may try to dominate the other. We really don't know what goes on. We only know what results.

In the lengthy sixth stage of transgenics, the zygote, now reimplanted in a surrogate mother, replicates itself billions of times until birth. Every cell will now contain the nonspecies gene. The creature is transgenic. The monster is born. There are, however, some Saturday night variations. The embryo may abort during gestation. Rarely do we know if this was due to some explosive result of the blind date we arranged or was just the result of random chromosomal incompatibilities and mutations. Sometimes the monster may be a mosaic. This means that some of its cells have copies of the injected gene and other cells have copies of the original gene. A mosaic creature occurs when the injected gene squeezes itself into the zygote without kicking out the previous occupant. One copy of this original and the somewhat chromosomally schizophrenic "Eve cell" may have the normal gene fragment, and the other copy may have the injected gene fragment. As a result, the birthed creature will end up with about half of its cells with one set of instructions and the other half with a slightly different set of instructions.

Assuming we have survived the darkness of gestation and the thunder of birth, a potentially transgenic life-form arrives on the seventh day of creation. Whether it really is transgenic depends on two further criteria: Does it express the transgenic characteristic biochemically and does it pass on the transgenic characteristic hereditarily? Many times scientists create an animal that is transgenic in its genes but does not actually do anything transgenic. There are several possible reasons for this. One of the most common is that there may be other genes, usually called "promoter genes" that are necessary for a particularly coded ability to express itself. For example, one set of genes may tell the body how to synthesize a necessary enzyme, while another set of genes tells the body how much to make of that enzyme. If the animal is transgenic only with the first set of genes, nothing will occur because the animal's biosystem does not know what actually to do with the instructions it has. There are also many kinds of promoter genes. For example, if a company wants to "milk its herd" for a rare human enzyme, it had better be sure that the goats are transgenic with promoter genes that direct the enzyme to be secreted in goat milk. If, instead, the enzyme is secreted only in blood, the herd will not have gained much economic value.

Even if the animal does the transgenic thing during its life—makes human blood, produces rare human proteins—scientists and biotech tycoons will not be satisfied until it produces transgenic offspring. The reason for this is that it is too expensive and chancy to create a single transgenic creature. At best, only about one of every ten attempts work out. The business of transgenics demands mass production, which means in this industry that the animal factory literally replicates itself. It was this need for transgenic mass production that led to the famous sheep and monkey cloning experiments of 1997. For as yet unknown reasons, only a few transgenic creations pass on their transgenics to their offspring. When this does happen, however, a new subspecies is founded, far beyond anything ever likely to occur in nature. Unlike animal husbandry or selective breeding, transgenic creationism does not gradually cull a naturally occurring trait into greater and greater prominence, like a greyhound's fleetness or a calico's fur. Transgenic creationism takes a brand-new trait and, overnight, zaps it into existence. And while today the brand-new traits already exist somewhere in the catalog of living species, that catalog is so vast that the traits may as well be created out of thin air. With the knowledge gained from the Human Genome Project, the thin air of today may become the soft chromosomal clay of tomorrow.

should we fear transgenics?

The very essence of the practice called transgenics sparks mental connections to concepts such as creatures and monsters. If it is common in nature, it is natural. If it is not, then we scream "monster"; we call it a concoction, a creation, a creature. The words "monster" and "creature" spark fear, mostly of the unknown. We should, however, overcome our innate fear of transgenics. The practice could very well save our lives.

One of the most exciting applications of transgenics is growing organs for transplantation into humans that will not be rejected by the body. These immune-friendly livers and hearts are called "xenografts," grafts from something alien, from something transgenic. In America alone, at least one person receives a heart transplant every day. In some sense, these transplant recipients are already transgenic, for part of their biology does not come from their genes. It is not a big logical step to accepting a heart from a transgenic pig.

In the transgenic world of tomorrow, no one would have to suffer or die due to the decay of some part of the anatomy. Transgenic "pharms" will supply hospitals with organs just as supermarkets are supplied with meat. With spare parts grown on the pharm, artificial joints made out of plastic, and new pharmaceuticals brewed from genomics, we are likely to see a major extension of the human life span. Furthermore, by drastically reducing the cost and increasing the accessibility of spare organs and rare drugs, transgenics will benefit people everywhere in the world, not just the rich. Transgenics is to biotechnology what mass production was to cars. But, just as the assembly line forced a reevaluation of our ethics of work, genomic engineering forces a reevaluation of our ethics of life. Nothing will bring the ethical issues to the foreground faster than the use of transgenics to enhance positive eugenics.

With the transgenic genie in our midst, it is inevitable that it will work its magic not only on animals, but on humans as well. With this reversal of how transgenics is practiced today, snippets of DNA will be applied not to a fertilized animal zygote, but to a human one. The DNA fragments we isolate may come from an animal, or perhaps from another human. The DNA fragments may offer general immunity from cancer, extraordinarily long life, or special visual, mental, or physical acuity. As in the seven days of transgenic creation outlined above, millions of copies will be made of the "golden genes" before they are injected into the kind of human zygotes found in fertility labs in nearly every major country today.

As with transgenic animals, the transgenic human embryos will replicate repeatedly until the "golden genes" are in every cell of their body. If the appropriate "promoter genes" are also properly grafted, the transgenic person's body should express with proteins and other biochemical building blocks the instructions contained within the alien genetic material. In other words, as newborns grow up, they should have general immunity from cancer, extraordinarily long life, special acuities, or whatever other characteristics were genetically encoded. The transgenic person could also be expected to pass at least some of these attributes on to their children "the old-fashioned way."

There are three kinds of fears about using transgenics for personal eugenics. On close analysis, however, none appears likely to eventuate. The fear of the mainstream genetics community is that mistakes will be made during the transgenic creation process, resulting in deformed or debilitated children. Some social historians and humanists are afraid that the new creations will consider themselves a "superrace," precipitating conflict between the genics and the transgenics, with the likely result of massive demographic death. Many theologians worry that transgenics will lead to a reverence for body over soul, with a consequent decline in morality.

Fear of tort lawyers will likely prevent the use of transgenics for personal eugenic enhancement until the risks of unintended harm have been reduced virtually to nil. If "monsters" are born, the geneticists will be sued for medical malpractice, the hospitals for permitting negligent mayhem, and the parents for wrongful life. It is unlikely that insurance companies would maintain malpractice insurance for medically unnecessary procedures that could result in huge tort liability. Without this kind of insurance, no one in a professional capacity would engage in human transgenics until the practice was assuredly safe.

If tort liability does not dissuade doctors from undertaking human transgenics, it will be because the procedure has become entirely safe, or the risk of harm is so small that malpractice insurance premiums are adequate to cover any lawsuits, or that malpractice lawsuits will not succeed. In either of the first two conditions, the fears of the mainstream genetics community would no longer be valid. This would not be the first time that the mainstream genetics community feared something one year, only to retract their fear a few years later. This is exactly what occurred with the original technology of splicing off DNA fragments, in recombinant DNA. Geneticists imposed a voluntary moratorium for one year and revoked it two years later.

Malpractice lawsuits might not succeed under three circumstances: (1) parents were able to waive both their and their offspring's right to sue as a condition of having transgenic assistance, (2) new laws were passed that limited the ability to sue for the results of transgenics, or (3) courts held that transgenic procedures were medically necessary activities for which no one can be liable unless there is negligence. All three of these circumstances rest on the assumption that, although transgenics occasionally caused undesired effects, most of the time it generated huge benefits. Parents would be most unlikely to ask for transgenic assistance unless they thought they had only a vanishingly small chance of having to care for an invalid for life, but were virtually certain of producing a child stronger, healthier, wiser than themselves. No legislature would pass a law limiting the right to sue for transgenic procedures unless there was a veritable groundswell of demand for the procedure. No such demand is likely if badly deformed babies appear on the evening news.

Finally, the courts have long realized that no medical treatment is risk-free—people die from allergic reactions to aspirin and from polio vaccinations. So long as a transgenic treatment is carried out with the reasonable standard of care practiced by the genetics community, it cannot be negligent and hence cannot be the basis of a tort lawsuit. If there are no reasonable standards, it means that the procedure is experimental. Here, the doctor will be insulated from a lawsuit so long as the patient knowingly consents to the experimental risk.

Of course, the real patient in a transgenics procedure is not yet born. How can the offspring consent? Both statutory and case law give parents an absolute power of consent for medical treatment of their children. And what if the transgenics treatment is not medical but cosmetic? Parents may face a heightened risk of lawsuit for unfavorable outcomes of cosmetic transgenics treatments. But the very risk of such a lawsuit will eliminate the possibility of such treatments until the chance of an unfavorable outcome is nil. The geneticists' fear of producing deformed babies is not based in reality, because tort liability will prevent human transgenics while the risks are large, and tort nonliability will insulate human transgenics once the risks become small.

The second fear of human transgenics is the humanist fear of a superrace. This is quite the opposite fear of geneticists, who generally doubt their ability to pull off such a feat, and fret instead over deformities, all the while creating supertomatoes, supercows, and superpigs. Nevertheless, the humanists look at a human history littered with peoples clinging to any observable—and often unobservable—difference among themselves, and using that difference to justify a higher status for themselves. The whites against the blacks. The Aryans against the Jews. The Serbs against the Bosnians. The list is almost endless.

The humanist fear of transgenics appears to be misplaced for two main reasons. One is that human fratricide is not due to differences, but due to a lack of mutual respect regardless of differences. A second reason is that transgenics is by its very nature a mass-market technology, the very antithesis of superrace elitism.

The bases of fratricide are made, not born. Even physically similar people fight among themselves. If human subgroups cling to any observable—or unobservable—difference among themselves to justify status, then logically, the creation of new transgenic differences among people will not stop or even affect the process of claiming status. The problem is not the existence of differences, but the mindset or philosophy that a person's status is in any way a function of their biology.

Indeed, transgenics will help convey the message that "biology is *not* destiny" by making it clear that biology is something that anyone can change from generation to generation. The philosophy of transgenics must be that each soul deserves equal respect, regardless of the species through which that soul is expressed. After all, it will be expected that from generation to generation, and within generations, the human genome will diversify and evolve. Transgenics will help generate respect for all people as individuals by making it clear that differential biology need not be god-given, and can be man-made.

The very profuseness of transgenics also militates against the rise of a superrace. Because DNA copies are virtually free, any gene sequence worth having is copied millions of times. The copies are easily made, easily stored, and easily transported. It is extraordinarily unlikely for a gene fragment to be scarce. Hence, if any gene sequence were recognized as being so desirable as to give one "superhuman" capabilities, that gene sequence would be distributed everywhere, in short order, at extremely low incremental cost.

Even if some producers succeeded in hiding their test tubes

of the superhuman gene sequences, they could not stay hidden for long. Each superhuman person would have those gene sequences expressed in every cell in every part of the body—in each hair of the head, each flake of skin, each drop of blood. A simple comparison of the standard human genome against a hair, skin cell, or blood drop from the superhuman would yield the superhuman gene sequences. Shortly thereafter, riflips would cleave the sequences, bacteria would make millions of copies, and hypodermic needles would be injecting it into fresh zygotes—always assuming that people would *want* to be superhuman.

On the other hand, if the superhumans were a real menace to humanity, their genetic uniqueness would be a terrible Achilles' heel. Biological weapons that targeted only those persons with the superhuman gene sequences would be highly specific weapons. If some future society had the transgenics to make menacing superhumans, it would also have the technology to bring them down.

In short, because of the plentitude and transparency inherent in genetic technology, there is little about transgenics that is likely to give rise to a superrace. Genetic differences are cheap to copy, hard to hide, and tempting to target. On the other hand, there is much in transgenics for humanists to celebrate. Gene fragments that might eliminate suffering due to many diseases could easily be replicated in quantities adequate to serve every new child on earth. And human culture will likely achieve new prominence as biology itself falls clearly within the realm of creative choice. Hardly the stuff of superracialism, transgenic creation is merely humanism, free of hubris and full of hope.

Finally, theologians fear transgenics as an exaltation of the body over the soul. They see a usurpation of the deity in the creation of life, a pollution of nature in the cross-fertilization of species, and a disrespect for moral development in the engineering of human traits. These criticisms are neither accurate nor valid.

All that transgenics involves, for personal eugenics, is an extrapolation for modern times of the age-old right of people to choose their mates and, thereby, shape the fate of their offspring. If the soul is inherent in human spirit, and not in the human body, then the construction of that human body from one or another set of materials cannot possibly affect the validity, or importance, of the human soul. Our souls are not affected by the transplantation of heart parts from swine or hip joints from silicone. So how could it be different if that swine heart grew within us from the start, or if that hip joint grew strong from an improved chromosome instruction set rather than a space shuttle spin-off? Indeed, if our souls transcend our bodies, our souls must transcend our genes.

It is also not plausibly consistent for theologians to oppose transgenics on the basis of the violation of some deity's natural order of things. Humanity has interfered with the natural order of things from time immemorial. The cultivation of crops, domestication of animals, and vaccination against disease are but three of many examples. No major religion opposes these rearrangements of God's world. For the same reasons, there is no need to fear transgenic arrangements of God's world. The forging of zygotes through transgenic creation no more usurps holy monopolies than does the forging of zygotes through romantic intimacy. Transgenics just brings the players to the party. It is something magically quite beyond us that turns on the music and breathes life.

Last, human attentiveness to transgenics implies no inattentiveness to moral, ethical, or spiritual development. Indeed, quite the opposite is more likely involved. The more people are concerned about life, the more they are likely to be concerned about living. Life may be a matter of personal health, but living is a matter of social contracts. We live in a society of people, and it is the moral code, the ethical principle, and the spiritual bond that guide us in getting along with others. The longer we have life, the more important it is to have a good one. Hence, life-enhancing transgenics is a catalyst, not a substitute, for moral development.

An interesting example of how transgenic possibilities focus attention on "meaning of life issues" normally the purview of religion was presented in the May 1996 issue of the *New England Journal of Medicine.* The *Journal* reported the identification of two "clock-genes" in a worm that, when mutated, can extend the worm's life span "by 5 to 6 times, the largest increase in life-span ever recorded in any organism." The staid science publication then asked, "How do we translate these findings to humans?" Assuming such genetic mutations could safely be induced in analogous human genes, or could be transgenically spliced from worm DNA onto human DNA, "they raise what may be the most profound question in biology: are senescence and death really immutable?"

We have seen in this chapter that the nature of our offspring is a subject of immense interest to humanity. A subjective field of more perfect offspring, eugenics, has arisen to attend to this human concern. There are two opposing kinds of eugenics—personal and social—and two ways to implement each—positive or negative. What personal eugenics hath—free choice—

social eugenics taketh away. And what social eugenics is—public control of reproductive choice—personal eugenics denies. There are a multitude of ways to positively promote eugenic outcomes or to negatively quash eugenic mistakes. This catalog of conception is about to be expanded into an entirely new dimension with the advent of transgenic creationism, the blending of genetic material from parents and others.

It appears that society is happiest, and most free from suffering, when social eugenics is kept well at bay and personal eugenics runs truly free. To accomplish this prime directive, in the Age of Genomics, will require new social understandings, new legal conventions, and new moral standards. Indeed, we must implement a new global credo, a bioethics of birth. For just as industrial technology led to the demise of the arranged marriage, genomic technology spells the end of the happenstance child. Protecting our personal eugenic freedom under these circumstances requires foresight, vigilance, and proactivity.

We have only a decade or less. The Human Genome Project will give our age-old quest for the perfect child its ultimate tool, as the Manhattan Project gave our age-old quest for weaponry the atomic tool. But genetic Hiroshimas must be avoided without killing off genetic medicine as well. After all, nuclear medicine serves us well. We either act now to discipline the fiery magic of genomics, or we will surely feel the burning pain of its uncontrolled flames. Hence, we must now turn to the specifics of how to master this new genie, of how to control our own code.

part III the bioethics of birth

The history of bioethics over the last two decades has been the story of the development of a secular ethic. Initially, individuals working from within particular religious traditions held the center of bioethical discussions. However, this focus was replaced by analyses that span traditions, including particular secular traditions. As a result, a special secular tradition that attempts to frame answers in terms of no particular tradition, but rather in ways open to rational individuals as such, has emerged. Bioethics is an element of a secular culture and the great-grandchild of the Enlightenment.

H. TRISTRAM ENGELHARDT, JR.
The Foundations of Bioethics, 1986

Bioethics is based on eudaemonism, a philosophical tenet that holds that human happiness is a moral obligation to be achieved from actions based on reason. Bioethics is our logical almanac for harvesting the seeds of sex, because we want our offspring to be happy and to live in a happier world. It was reason, not doctrine, that opened up the genome for human exploration, and it must be reason, not doctrine, that helps us navigate our way. The explosion of human genome technology in our midst can be managed eudaemonically with four fundamental bioethical principles. In summary terms, these are:

1. The human genome belongs to us all worldwide.
2. Each of us has an absolute right intentionally to create new versions of the genome.
3. Society has a right to help prevent unintentionally created versions of the genome.
4. The genome cannot be a basis for discrimination of any kind.

These four bioethical dictates, extrapolated fully and applied faithfully, will ensure that the Human Genome Project spawns a renaissance of life and health rather than a nightmare of holocausts and death.

In some way or another, these bioethical dictates will offend the moral sensibilities of most religious value systems. However, such an outcome is intrinsic to bioethics. Only by living under a theocracy can we be sure of avoiding offense to at least one religious system. The beauty of bioethics is that it justifies any offense caused from "naked rationality," rather than by resorting to a premise of one or another messianic doctrine. In other words, with bioethics, even the assumptions are always put to an acid test—will these assumptions lead to a happier world? Usually this acid test results in an assumption that individuals should be let alone except for such minimal intrusions as are needed for the common good.

In a diverse, multicultural society, the bioethical avoidance of all religious truths will appear to be more fair across different belief systems than will the idiosyncratic appeals of each religious group. Nevertheless, reasoning from religious truths will often coincide with bioethical conclusions. Just as every religion "interpreted away" forcible conversions of faith in the higher interests of a more peaceable world, the time has now come for religion to "reason away" archaic views of procreation in the higher interest of a more healthy world. It is in this vein that physician–philosopher H. Tristram Englehardt, Jr., refers to bioethics as the "lingua franca of a world concerned with health care but not possessing a common ethical viewpoint."

chapter 6 sharing our genome: the fabric of life

A species is a group of populations whose individual members would, if given the opportunity, interbreed with individuals of other populations of that group. Thus all human populations, no matter how different they look, belong to the same species because they do interbreed and have interbred whenever they have encountered each other. . . . On average there's 0.2 percent difference in genetic material between any two randomly chosen people on Earth.

JARED DIAMOND
"Race Without Color," *1994*

We must begin by recognizing that the human genome is the common heritage of all humanity. This means that no one can own the human genome and no one can pollute the human genome, but everyone can use it, including biotechnology businesses. The human genome is not the first thing to be considered as humanity's common heritage. Since the 1960s, this is the way diplomats have looked at outer space and the high seas. As a result, everyone enjoys the benefits of satellite communications and transoceanic trade. There are no customs gates or tariff tolls in orbit.

Of course, it takes some mental adjusting to think of a microscopic genome as the common heritage of humanity. But its size turns out to be irrelevant. What matters most is that all humanity must share the human genome, as it must share the oceans, heavens, and skies. What is important is that national property rights in the genome would cause conflict among countries, as such rights have in the past with land and as has, fortunately, been avoided with deep space. What is essential is that we pass on the human genome to future generations, not consume it like a Texas oil well or pollute it like a Love Canal. The legal concept for sharing a resource among all nations, without rights of pollution or depletion, is the common heritage principle. It must be our first bioethical principle in the Age of Genomics.

A common heritage means that no person, business, or government can prevent others from making use of it. For example, it would be wrong to patent parts of the human genome, because such a patent would prevent others from using it. As discussed in Chapter 2, the U.S. government tried to patent the parts of the human genome that describe how the brain works. Although the government changed its mind under European political pressure, there is still no legal determination of the patentability of human life. This possibility must be precluded absolutely, for patents on the genome are just as wrong as blockades on the seas.

There is some clamor from biotechnology companies that without patent rights they have no incentive to undertake the necessary research to develop genomic pharmaceuticals. Their concern is misplaced. Even if the genome itself cannot be owned, companies can still obtain proprietary rights in what they discover using the genome. For example, specific transgenic formulations that produce specific results should be patentable, by analogy to the fact that, although no country can claim ownership of outer space, proprietary rights can certainly be claimed to the use of a particular piece of outer space by a particular satellite or space station. That is how we are able to enjoy CNN and MTV today. The grafting of an MTV satellite onto a piece of outer space is like a transgenic creation. Without the MTV satellite, or without the transplanted genes, there would be only an unownable, common heritage piece of outer space or inner genomics. With the grafting, there are proprietary economic rights equal to the extent of the actual graft.

In the view of the European Union, which is grappling with this same problem, the human genome "itself" may not be patented, but products derived from it may be. In a similar vein, countries may not own the ocean floor, but they may own any minerals they extract from the ocean floor. It is fundamental to any common heritage resource that the resource itself cannot be owned, but that beneficial things extracted from the resource can be. In this way ownership of the resource itself does not become a source of conflict, yet organizations still have incentives to develop it commercially.

There may be further concerns that even permitting the patenting of exploited pieces of the human genome somehow denigrates human life or obstructs humanity's access to the genome. Neither of these concerns appears valid. First, it is not a life that is being patented, but the specific manner in which that life was created. For example, suppose a company patented a gene for superb night vision as part of a technique for splicing the gene onto a zygote. The visually acute individuals who have eventually benefited from this patent are in no way owned by the company. Second, patenting the method of producing one kind of person in no way diminishes the humanity of others. Our humanity, for example, is not affected by the hospital in which we were born, or the parents who conceived us. Humanity is equally descriptive of all birthed persons, regardless of how they were conceived and born. In essence, patenting transgenic methods is really no different from patenting devices that assist childbirth decisions, such as ultrasound machines.

Finally, we must consider how the common heritage principle's ban on pollution affects something so small, and yet so ubiquitous, as the human genome. Pollution is a general reduction in the value of some resource. It may be smoke in the air, waste in the water, or noise in the quiet. Pollution need not affect the entire resource, but it must affect more than one's own backyard. A general reduction in the value of the genomic resource must affect many people's bodies and the genomes within them. International law already covers most pollutants that come to mind. Environmental poisons that cause mutations in the genome resource are certainly a pollutant to proscribe. Similarly, bacteriological weapons that affect any subset of the human genome—for example, just one particular ethnic group or gene pool—would definitely be a pollutant to prevent.

In a 1996 "Report on the Human Genome," UNESCO argues that hybridization with other species, such as transgenics enables, is a pollution of the human genome, because it challenges the "integrity of the human species, as a value in its own right." This view is mistaken because it presumes a "genism" viewpoint that is quite analogous to racism. The blending of nonhuman genes into humans no more reduces the worth of the human genome than the children of mixed-ethnicity parents reduce the worth of either parents' ethnic group. Contrary to UNESCO, hybridization enriches genomic diversity; the integrity of the human species rests on social solidarity, not genetic consistency.

The human genome is not for sale. Not for lease. Just for use. No fencing. No dumping. Only sharing. This is the bioethical essence of the principle of treating the human genome as the common heritage of humanity. Adoption and implementation of this principle alone would obviate some of the Human Genome Project's worst-possible horrors and bring some of its most beneficial uses to the fore.

In her landmark 1995 *University of Chicago Law Review* article, "The Genetic Tie," Dorothy Roberts has aptly observed that "in this society, perhaps the most significant trait passed from parents to child is race." She supports this view with a long list of judicial decisions in which inheritance, child custody, and other rights were decided in a way that usually ensured racial separation and almost always ensured "the paramount objective of keeping the white bloodline free from Black contaminations." For examples, as recently as the 1960s, half the American states forbade interracial marriage; until 1980, some states legally prohibited African Americans from adopting Euro-American children; and even today, "race" is routinely used by judges in settling family law disputes, usually in favor of "racial purity" outcomes. Consequently, Roberts concludes that simply grafting reproductive technology onto a racist society will only make it more racist—exploited women of color serving as sterilized surrogate gestation mothers for Euro-American professionals, growing gaps between the health of mostly well-off genomically enhanced whites and not so fortunate minorities, and advertising of any genetic manipulation that enhances "whiteness." She describes this last indicium of racism tellingly: "New reproductive technologies are so popular in American culture not simply because of the value placed on the genetic tie, but because of the value placed on the white genetic tie. . . . It is hard to imagine a multimillion dollar industry designed to create Black children."

Roberts is right. An imperative concomitant to the bioethics of birth is for much greater effort to be made to eliminate racism in society. A starting point is government funding of full genomic health care, and related educational outreach efforts, for people living at or below the poverty line. Having made this investment in procreative choice and genomic health, the government needs to follow through with a revitalized commitment to equal educational opportunity, including community worker outreach to those segments of socity that have been left out of the economic mainstream.

Race is an invention; peoples are a mélange. Scientifically, there are not races, just countless graduations of phenotypic and biochemical human characteristics. Hundreds of years have been spent ideologically propping up separate racial categories as a justification for the embarrassing dilemma of slavery and its aftermath in a society nominally dedicated to "liberty and justice for all." While the reality of racism causes excruciating pain to millions under agricultural and industrial economies, it can cause absolute democidal annihilation in a postindustrial biotech world. Accordingly, we who invented race must disinvent it as rapidly as possible and must commit to spending the needed billions of dollars in defense of our nation's diversity. Sensible outreach to the oppressed sectors of society, practical sharing of the full panoply of biotechnology regardless of income, and vigorous improvements in education and social environments in every part of our cities are the surest ways to redefine race as the mélange of all human genomes that it, in truth of scientific fact, actually is.

Until an appropriately funded antiracism effort is in place, we must rely on image makers and opinion leaders to reinvent race as the human mélange. The Age of Genomics will not stand still. And neither can we. Every person must pull for our common genome and agitate for its liberation from racial stereotypes and oppressive restraints. Such agitation means celebrating the contributions of all our demographic peoples, especially the unfairly maligned peoples of color. It means promoting genomically mixed families, with diverstiy going in all directions. And it means working for national legislation that makes reparations for the ravages of racism, starting with payment for full reproductive options, health care coverage, and educational initiatives for all.

Social solidarity is the realization of the human genome. It is humanity's common heritage in practice. Today our solidarity is shaky, and our common heritage is in dispute. This must be rectified before the cloak drops off the human genome, and our social fabric gets torn beyond repair. Our challenge is to ensure the victory of science over sterotypes and of justice over prejudice. Only with a commitment to social solidarity, and pride in our multicultural heritage, can the fabric of human genomics be the universal healing blanket that is its highest calling.

chapter 7 expressing our genome: intentional life

Having babies . . . may be the most reasonable and available choice, a natural outcome of all the forces in their lives, in which avenues for self-definition and expression other than mothering are largely absent.

NANCY DUBLER AND CAROL LEVINE
"The Reproductive Choices of HIV-Infected Women," *1990*

The intentional creation of offspring can be the most vital, most meaningful, and most essential part of a person's freedom of expression. People who never write and rarely read books, seldom give and rarely hear speeches, nevertheless express their deepest personal convictions through the love of a child. As with other constitutionally protected forms of expression, the rule should be simple: Congress shall pass no law abridging the freedom of birth (speech). By preserving personal, positive eugenics as an absolute right, we ensure that all social eugenics is an absolute wrong.

The freedom of expression rationale for personal genomics is useful in two different ways. First, it lets us understand the absolute importance of personal eugenic expression. It is analogous to freedom of speech—the touchstone of a democratic society—because intentional childbirth is the ultimate expression of our theological, philosophical, or political will. Second, freedom of expression gives us an analytic framework of "First Amendment law" as a set of analogous guidelines for applying this penultimate bioethic of birth to everyday life.

The right to procreate is a fundamental human right: to extend ourselves through time, so that whatever cannot be accomplished by us personally may be accomplished by our children. As the Bambara say, "The only remedy for death is a child." Since 1947, the World Population Plan of Action, based on the UN World Population Conference, has declared: "All couples and individuals have the basic right to decide freely and responsibly the number and spacing of their children and to have the information, education, and means to do so."

From their earliest days, we transfer to our offspring our beliefs, attitudes, and values, expressing ourselves, our souls, in the most intimate, immediate, intrinsic way possible. Procreation is the epitome of expression, copying body to body, mind to mind, soul to soul. It is at once religious, personal, philosophical, and political. America's founding patriots and Supreme Court justices have made clear that the stringent First Amendment constitutional guarantees protect expression that is political, philosophical, or religious. Given this mandate, there is nothing more deserving of freedom from interference than our expression of birth—our eminently philosophical, religious, and/or political decision to have a child.

To be clear, the expression inherent in procreation is not the genomic code itself, but instead is the process of parenting. Parental expression can be accomplished just as powerfully with adopted children as with one's own biological offspring. As reproductive law scholar Dorothy Roberts has observed, "Cultural forces dictate what powers we bestow upon these particles passed from parent to child." Sometimes these forces virtually make the medium the message, although usually the genetic endowment is at most a carefully thought-out template for transcribing one's soul.

Freedom of personal eugenics, like freedom of speech, has a number of immediate implications:

Freedom of Personal Eugenic Expression	*Freedom of Personal Verbal Expression*
From a government entity's order for sterilization or abortion	From a government entity's imposition of a priori restraint on a publication
The right to express oneself in whatever offspring one wants	The right to express oneself in whatever words one wants
Acceptance of consequences for "wrongful life"	Acceptance of consequences for "wrongful speech"
No social eugenics	No government limits
The right to use regular zygotes or hybrid transgenics	The right to use words, multi-media, or Internet
We must, nevertheless, take personal responsibility for our actions: financially care for our children, develop them, and have tort liability for them.	We must, nevertheless, take personal responsibility for our actions: pay for our speech, create our speech, and accept libel/slander liability.

Personal eugenic freedom is very similar to personal religious freedom in that both rights are absolute. The Supreme Court has ruled that even strange and unpopular theologies are entitled to the religious freedom protection of the First Amendment. Similarly, even mothers certain to bear children without a brain cannot be forced to abort their infants. The reasoning is the same in both cases: Any line drawn between acceptable and unacceptable religions, or children, is bound to be arbitrary, and therefore slippery, and therefore possibly capable of undermining the entire fundamental freedom. It is better to draw no lines at all, to leave choice of children, like choice of religion, up to the people expressing themselves.

Freedom from interference means no interference, direct or indirect. The government could practice social eugenics in all sorts of indirect ways: for example, only health care premiums for genetically counseled and selectively fertilized offspring could be deemed tax deductible, by reasoning that such children will cost less in total national health care expense. But such treatment would be as wrong and as unconstitutional as providing indirect tax benefits only to conservative forms of speech as a means of enhancing social stability.

The absolute right to have offspring, and the absolute bar on government infringement of this right, must also extend to transgenic creation. In a similar vein, the courts have increasingly held that the First Amendment guarantees of free speech apply equally well to satellite broadcasting, multimedia, and the Internet—all technologies far beyond the conception of our Constitution's authors. Transgenic creation is to plain old-fashioned coital conception as the Internet is to town hall speech or hand-printed books.

Some critics will worry that transgenic creation may enable a parade of horribles to enter our world. Let's consider what kind of children parents might transgenically create:

- An Orthodox Jewish woman who has her heart set on producing a very East European–looking child, sets out to marry an Ashkenazi rabbi's son, using transgenics to replace the part of their sperm and egg cell DNA that codes for Tay-Sachs disease. They thereby eliminate Tay-Sachs disease from their germ line rather than aborting embryos that test positive and keeping embryos with only one of the invariably fatal two Tay-Sachs genes.

- A Hollywood film star has his heart set on a female child as deep melanined as some Africans, but with long blonde hair and blue eyes. Instead of looking around the planet for the mate who might help him achieve his desired genetic expression, he purchases an anonymous egg cell, uses a transgenic sequence from an African for epidermal melanin and transgenic sequences from a Swede for hair and eyes, and has the child gestated by a surrogate mother. A female child is guaranteed because only sperm carrying an X chromosome are transgenically modified. She becomes a famous model due, in part, to her uniquely striking looks.
- An Alpine couple want children even more physically rugged than they. They could simply purchase sperm that comes from a donor with an ancient heritage of adaptation to rugged climates, such as an Andes Indian. Instead, they find scientists who had isolated gene sequences from certain animals that are very similar to human gene sequences but with better metabolic processes in sharply reduced temperatures or under significantly reduced oxygen, as occurs at high altitudes. They have their sperm and egg fertilized in a petri dish, transgenically modified, and reimplanted in

the mother's uterus for gestation. The child becomes a world record-holder in Himalayan mountain climbing.
- A New Zealand couple have the misfortune to live under the worst part of the ozone hole, and skin cancer plagues nearly all their friends. Gene sequences from a rare desert animal have been found to virtually eliminate the risk of cancer. The couple could have used sperm from their few friends who were less susceptible to skin cancer, but decided instead to fertilize their egg and sperm cells in a petri dish and have the animal gene sequence added. The transgenic offspring acts and looks perfectly normal but never gets skin cancer.
- A transsexual Chinese couple want to have hermaphrodite children so that they can enjoy the pleasures of both sexes. The portion of the gene sequence that turns on male sexual characteristics is added to a sperm carrying only an X chromosome. The resultant child has both male and female reproductive tracts.

Many of these possibilities may seem bizarre to us—new races, hybrid animal/humans (!), or even new sexes(!!)—yet they can be created today in other ways. Many Jewish parents of East European origin currently abort embryos that are certain to carry Tay-Sachs disease. Hollywood couples frequently seek out mates with the looks of their offspring in mind, and then use plastic surgery to get it just right later. Other parents select mates with the intention of passing on their specific physiological or intellectual traits to their children. This is the reason sperm banks provide the "looks and thinks" information in the table in Chapter 3. For example, sperm bank semen donated by a "poet" might, some parents think, lead to a more creative child. And many traits are simply created medically—hormonally enhanced muscles, laser-enhanced eyesight, surgically augmented breasts.

While the technology for human transgenics is not yet available, it is coming rapidly enough that some people are asking for a moratorium on such activities. The biological expressions described above are mundane compared to some constitutionally protected forms of expression now appearing in religion, books, and film. None of these transgenic creations is likely to spell the end of life on earth, but that might be the consequence if we permit governments to tell us what kind of children we can or cannot have.

The risks that parents using transgenic creationism will bring into life ghastly, wrought in deformed agony, or totally insane newborns are no greater than today's while making babies the "old-fashioned way." And couples are totally within their rights not to abort a child that doctors are certain will be a "freak." They are totally within their rights to conceive a child with a 1 in 4, 1 in 2, or even 1 in 1 chance of having a serious, debilitating congenital disease. So, if these rights exist without transgenics—and they must lest Big Mother starts to say what kind of children we can or cannot have—then, why shouldn't they exist with transgenics? There is even more reason for transgenic freedom, however, because the overwhelming probability is that the use of the technology will make life better for the offspring.

A second reason not to fear neonatally cruel transgenic outcomes is that tort law provides a fairly solid remedy, with the threat of multimillion-dollar lawsuits serving as a powerful disincentive. The remedy, called "wrongful life" or "wrongful birth," allows children to sue doctors, hospitals, and even their parents for bringing them into the world in such disastrous condition that they would prefer to be dead. For example, doctors have been held liable for millions of dollars in damages for failing to advise a mother that she was carrying a Tay-Sachs child that could be aborted. The threat of bankrupting damages for bringing an "elephant man" into existence will limit the use of transgenics to only the most assuredly safe procedures.

The use of tort liability to manage transgenic creationism does not contravene the principle of personal positive eugenic freedom. Notwithstanding freedom of religion and speech, people are responsible for the consequences of their expression. A cult is free to preach a message of apocalypse, but if cult members commit murder they will be held to account for it. A newspaper is free to derogate the reputation of another person. But if what the newspaper publishes amounts to libel, it will be monetarily responsible for the damage caused. Criminal and civil responsibility for the untoward consequences of one's expression is not inconsistent with one's basic right to unfettered expression. We should be free to play with the genes of our offspring, but if our play produces harm, we must be ready to accept financial as well as moral responsibility for the consequences of our actions.

Finally, it might be asked why society should have to pay, through health care costs, for the "irresponsible" procreative choices of certain individuals? If people mate in a genetically "unhealthy" manner, fail to abort "unhealthy" children, or transgenically produce "basket cases," why should society have to pay for their mistakes? Even by analogy to other kinds of freedom of expression, society does not have to provide a job to the Communist pamphleteer who becomes an outcast or a new house to the disenchanted former Krishna follower who gave her last one away. The answers here are simple. First, health care is like national defense—it is a duty owed to all citizens without regard to whether we believe they deserve it or not. We cannot shirk defending against foreign attack the children of even the most radical pamphleteer, deserter, or traitor. Nor should we shirk our health care obligation to the children of the wrist-slasher, the crack addict, or the gene meddler.

A second reason not to let economics distort our bioethics is that the cost of health care, even for "irresponsible" reproductive decisions, is a small price to pay to ensure our own reproductive freedom. It might be cheaper to suppress inflammatory speech, and thus not risk terrorist attacks, but that would undermine the principles of the republic we seek to defend. Similarly, it might be cheaper to suppress "unhealthy" births, and thus not raise health care costs. But that, too, would undermine the principles of the republic we seek to defend. If we start denying health care to some other people's genomes, we may soon find it denied to some of our own genomes. Few of us would have won America's "fitter family" contests of the 1920s so admired by the founders of the Third Reich.

A third reason to prohibit the threat of the withdrawal of health insurance from transgenic creationists is that it amounts to punishing its victims rather than their parents. Children are not the "irresponsible" parties, so why should they be denied any necessary health care? We should be outraged over proposals to make a sick child pay for the sins of his or her parents.

A final and very fundamental reason that society should pay

for the procreative decisions of its members is to keep genomic technology from becoming an amplifier of preexisting socioeconomic class differences. We hold dear the notion that in a capitalist or mixed economy everyone has a fair chance of material success. We owe the newborn at least the birthright of a level biological playing field, mediated by parental judgment, which, in turn, should be appropriately informed by knowledgeable peer-respected community workers. In other words, *class* advantages in genomics make a mockery of equal opportunity. There will be plenty of class advantages after birth. Society owes itself the responsibility of genomic fairness before birth. This much is essential to a meritocracy.

Accordingly, it is not enough that parents be permitted to use personal eugenics techniques. That, plus laissez-faire, is a de facto decision that the rich will get better and the poor will get worse. Instead, the promise of personal eugenic biotechnology can be realized only if the government empowers all of its citizens to access these techniques. Such empowerment, through legislative mandate and practical implementation, is fully justified as part of a country's social contract. Like the telephone network and the highway, personal eugenic technology must be available in practice to all as the price of being available to some. "Absent such a policy," writes feminist scholar Adrienne Asch on behalf of the feminist bioethics community,

> Genetic disorders will acquire even more stigma than now accrues to people who have disabilities. Thus, despite our qualms about much of the philosophy behind the Genome Project and our reluctance to suggest anything that will encourage the overselling of genetic ser-

> vices, we believe that equal access to these services is an important vehicle for preserving community by minimizing the harms that grow out of the basic inequality in the distribution of human characteristics.

In summary, the second bioethic of birth is our absolute right of intentional procreative expression. Its corollary is a ban on any government infringement of the right to express ourselves through our offspring. Every right has its associated responsibilities, and the right to personal positive eugenic expression is no different. If we cause harm, we can be held to account for it. Whatever our form of expression—be it with words, with gods, or with genes—we must be prepared to live with the consequences for as long as we live. Indeed, that is why people express themselves politically, philosophically, or theologically in the first place—to change the world.

chapter 8 controlling our genome: unintentional life

In both Ghana and Kenya . . . about 40 percent of married teenagers who have had children said their first pregnancies were unintended; among unmarried teenagers the proportion of unintended births rose to 58 percent in Ghana and 77 percent in Kenya.

WORLD BANK
World Development Report, 1993

More than 50 percent of the pregnancies among American women are unintended—one-half of these are terminated by abortion. 11 percent of American abortions are obtained by women whose household income is greater than $50,000.

ALAN GUTTMACHER INSTITUTE
1994

The third bioethic of birth considers unintentional genomes, such as those that force an unintentional pregnancy, to be conditions of disease rather than of expressed life. Although society may not prevent the expression of life, it does have a well-accepted role in helping to prevent the occurrence and growth of disease. The issue is no longer prochoice or prolife. It is, instead, one of creating laws to end the plague of genomic disease: unintentional pregnancies. Only within the last ten years has biotechnology made this possible. It is now up to us to make that biotechnology work for us, healing the wounds created over abortion, and ending the death, drudgery, and debasement of forced pregnancy.

It is, on first consideration, difficult to look at pregnancy as disease, or to think that any biological condition is or is not disease depending on our frame of mind—our intentions. After all, for thousands of years we have been taught that women are *supposed* to be pregnant. How can pregnancy ever be a disease if that's the way women are supposed to be?

The *Encyclopedia Brittanica* in its 1991 edition observes that: "disease is commonly considered to be a departure from the normal physiological state of a living organism sufficient to produce overt signs, or symptoms. The concept applies to the mental, as well as the physical, state of the organism." The key elements of the definition of disease appear to be "departure from normal," "overt symptoms," and applicability "to the mental as well as the physical." So, based on this typical medical-community consensus, we can begin to answer two of our questions: Disease may indeed depend on a person's mental state, and pregnancy may indeed be disease since it is obviously a departure from "the normal physiological state."

Now some people may take issue with the claim that pregnancy is a departure from "the normal physiological state." Probably most such people have never been pregnant. It is true, nevertheless, that for most of human history women were, essentially, always pregnant. Indeed, no less an expert than Martin Luther observed, "God formed her body to belong to a man, to have and rear children. . . . Let them bear children until they die of it; that is what they are for." In many parts of the world, women still labor under that advice.

But times have changed, and pregnancy is no longer the "normal physiological state" for women—if it ever was. Accordingly, what may not have been a state of disease in the past, is definitely a disease today. Nor would this be the first disease that arose as a consequence of human evolution. Normal conditions for barbarians are not usually normal conditions for suburbanites.

Now, although pregnancy can clearly be a state of disease, there remains the issue of mental state. As noted above, disease can depend both on physiological and mental status. The psychiatric community has addressed this issue in their definition of mental disease. According to the American Psychiatric Association's standard *Diagnostic and Statistical Manual-IV* (1994), three conditions must be met for mental disease to be present:

1. There is some pain, impairment in one or more important areas of functioning, or significantly increased risk of death, pain, disability or loss of freedom [pregnancy clearly meets these conditions—Third World women have a 1 in 20 chance of dying from each pregnancy, and all pregnant women are to some extent disabled].
2. The afflictions described above are not expected as a result of a specific event [for example, the *decision* to have sex, not to become pregnant],
3. The afflictions have some underlying cause in the person, and are not merely due to society's "dis-ease" with the person [the physical changes of pregnancy clearly meet this condition].

Accordingly, the bodily changes that occur as part of an unanticipated pregnancy reflect a condition of disease, not of life. If a person adapts to the unanticipated pregnancy—that is, decides that she wants it—then there no longer is a disease. Similarly, if persons adapt to their blindness, deafness, or paralysis, they are not disabled, they are differently abled. Disease can, indeed, be very much a matter of state of mind. But if women do not want their pregnancy, they are diseased and have a right to be cured. They have a right to an abortion.

Pursuant to the third bioethic of birth, society should first make maximum efforts to prevent genomic disease, and then be empowered to cure any outbreaks that nevertheless occur. In practice, in the Age of Genomics, this means that we must move toward innoculating people against unintended pregnancy. Biotechnology has given us the ability to do this with a new concept that may be called "inocuseeding." Universal inocuseeding means the banking of all men's semen accompanied by lifelong automatic contraception. While biotechnology is slowly developing vaccine-like contraceptives, including many studies of safe synthetic analogs of the antispermatogenic cottonseed extract gossypol, a technique available today is the simple vasectomy. The process is inoculative because a small harm is inflicted (the contraceptive vaccination or vasectomy) to prevent the possibility of a much larger harm (unintended genomes). Inocuseeding is not sterilization because it never occurs without simultaneous sperm banking.

With future technology—as near as a few years away—a safe injection or oral contraceptive may obviate the vasectomy. Before such an easily administered technology arrives, however, it is important to establish the second bioethical prinicple of birth, the freedom of procreation.

Technology now allows us to view pregnancy as either health, if intended, or disease, if not. As always, society should strive to maximize the health of its citizens, through inoculations and other actions. The Constitution of the World Health Organization (WHO) defines "health" as "a state of complete physical, mental and social well-being." An undesired pregnancy meets none of these criteria of well-being, and more than half of the pregnancies in the world are undesired. This means that society should be taking more aggressive steps to enhance the health of its female populace. Inocuseeding is the most effective and fair way to ensure that all pregnancies are intended. The age of happenstance children is nearing an end.

preventing disease

Of course, the mere mention of "vasectomy" sends shivers up many a male spine (although, in the United States, the National Center for Health Statistics reported in 1994 that 30 percent of men and 50 percent of women are sterilized by age forty-five). But this dread is entirely the result of the way vasectomies have been handled and the absence of any fair and reasonable bioethics of birth. Taking away a person's right to procreate is awful in that it violates our second bioethic of birth. It silences a parent's voice. But inocuseeding accompanied by the bioethics of birth is as reasonable and as necessary as universal vaccinations or safeguarding one's life savings in a vault. And it should be remembered that other kinds of inoculation have historically faced resistance. In William McNeill's 1977 book *Plagues and Peoples,* the story of popular resistance to smallpox inoculation is told, with a consequent huge toll of death and misery, because "opponents criticized the practice as an interference with God's will."

Vasectomizing procreative persons without banking their sperm is just plain wrong, foolish, and unnecessary. It is wrong because it may deny them the right to express themselves in offspring. It is foolish because it confuses personal eugenics with social eugenics. And it is unnecessary because, accomplished properly, sperm can be stored for a small cost that is overwhelmingly offset by the economic savings of reduced social expenditures for unintended genomes. The cost of inocuseeding, like the cost of all vaccination and inoculation programs, is a mere pittance to pay to avoid the agony of disease, in this case suffered by women from unintended, undesired pregnancies.

Sperm banking today is the process of taking sperm from masturbation, freezing it in one or more small plastic containers, labeling the containers, and storing them at very low temperatures. Sperm banked in this fashion has a half-life of more than a thousand years; in other words, half the sperm will still be "good" after a thousand years. Since each masturbated portion of semen contains well over a million sperm cells, there is nothing to worry about from the standpoint of quantity. As far as quality goes, it is no worse thawing out than it was coming in.

Banked sperm is used by sending it to the place of fertilization in a frozen pack. After removal from its pack, it thaws in a few minutes and is ready to fertilize an egg cell. This can be done by inserting the sperm directly into the uterus during a time of likely ovulation. While experts used to advise women to inseminate themselves up to two days after ovulation, the National Institutes of Health changed that advice in early 1996. New studies show that almost 90 percent of pregnancies result from insemination occurring within six days *prior* to ovulation.

The "artificial fertilization" process works about as reliably as the "old-fashioned way." While the process may not seem as "romantic" as the ancient "pump and squirt" approach, neither is buying plastic-wrapped, pasteurized meat in the frozen food section as "romantic" as eating a fresh-killed, blood-soaked, fly-covered animal. But people do evolve. And many thousands of couples have developed romantic home rituals for artificial insemination.

In the process of sperm banking, some "romanticism" is possibly traded for the benefits of ensuring that all conception is intentional and that all conception can benefit from predictive genetic diagnostics. It is really questionable whether romanticism is traded at all—artificial fertilization imposes no restriction on the frequency of lovemaking and eliminates entirely the need for coitus interruptus, "rhythm" calculations, birth control devices, and, for couples in long-term committed relationships, condoms.

Beyond romanticism, it is important to ensure that fertilization with banked sperm proceeds in a manner that respects the woman's body and soul. Her body's mystical signals are needed to guide the semen, nurture the zygote, and carry the offspring. Society is making a major invasion of a woman's body by asking her to take a lead role in the sociotechnological process of reproduction. We owe her every deference to her natural body cycles and eschewal of any kind of patriarchal insistence that she conceive "on demand" or "on schedule." We should be very reluctant to induce her conception with artificial drugs.

Obstetricians trained in women's studies are certainly capable of developing a feminist reprotech process, including educational materials on insemination, peer-counseling help lines, and in-person facilitators. There is nothing intrinsic to reprotech that need result in women's becoming what Gina

Cormea calls, in a different context, "Mother Machines." Indeed, the inocuseeding process glorifies the decision to become pregnant, and is thus likely to treat as special and spiritual its concomitant fertilization and nurturing rituals.

Sperm banking can occur as part of a government-sponsored inocuseeding program that includes boys during the year or two immediately after puberty. Safeguards are necessary to ensure that sperm is recovered prior to the boy's inoculative vaccination or vasectomy, and that the sperm is banked in two separate locations. Nevertheless, in the event of errors, there is no great problem, because vasectomies generally can be reversed with minor surgery.

Should inocuseeding via vasectomy be politically unachievable, the specter of it should be used to prod the government to fund a less-invasive universal contraception and seed-banking system. Research on safe, universal contraception needs to be accelerated because women are carrying an unprecedented and growing load of undesired pregnancy and abortion guilt-trips. This is not fair to women, to children, or to society's respect for life. The politicians must either give us inocuseeding via vasectomy and sperm banking or give us a massive research program toward an equally effective alternative.

Once boys are universally inocuseeded, the risk of unintended pregnancies will drop close to zero. Accordingly, the incidence of abortions also will drop close to zero. This will be a radical change from today when, even in the United States, studies repeatedly show that over half of all pregnancies are unintended. That's a lot of disease. The debate between choice and life will have been mooted by technology because, virtually always, life will be a choice, not an accident.

Inocuseeding must be accompanied by the bioethics of birth.

It is not reasonable to expect people to participate in a sperm-banking program if their access to their sperm is conditional on anything whatsoever. Simply put, banked sperm must be available on demand. If the government or sperm-banking authority were to impose any kind of restrictions on access to the sperm, then an incipient form of social eugenics would be in place. Whenever access restrictions are in place, some people get access and others do not. That is tantamount to regulating who may reproduce and who may not. Nothing of this sort is permissible under the bioethics of birth.

It is reasonable for the sperm-banking companies, or other entities, to offer confidential genetic testing services on banked sperm, removed eggs, or fertilized zygotes. This information would provide the owners of these seeds with confidential information, to the best of science's current knowledge, about the genetic propensities of their offspring. For example, Robert Windton, the British pioneer of in vitro fertilization (IVF), has begun to offer "pre-implantation diagnosis" to his patients at England's prestigious Hammersmith Hospital. Once petri dish–created embryos reach the 8- or 16-cell stage, Winston clips off a cell of each embryo and scans its DNA for genes that give a high probability of developing cancer by midlife. (His first patients carried a familial adenomatous polyposis gene, which leads to bowel cancer, a condition the parents preferred to leave in the petri dish.) Parents are then given the choice of which embryo to implant, each of which has its own genetic disease-proclivity profile.

No matter what gene-screening data are given to parents, it must be absolutely impermissible for this information to be used in any way to influence or affect the parents' choice to have offspring. The decision to have offspring, and if so with what characteristics, is the unassailable right of the parents. It is the essence of positive personal eugenics. It is freedom of expression, the second bioethic of birth.

Recalling our analogy to the First Amendment protections the U.S. Constitution offers freedom of expression, over the years the Supreme Court has fashioned what are called "time, place, and manner" restrictions. These qualifications of freedom of expression are never allowed to prevent any particular message or speech or religion. They are, however, allowed to establish reasonable "venue" limits as to when, where, and how the expression may occur. For example, anyone may drive around with a sound truck blasting a political message—but not between midnight and dawn. Anyone may march or demonstrate—but not on an airport runway. Bona fide religious ministers may practice ritual animal sacrifices—but not with the neighbor's dog.

In a similar vein, while the bioethics of birth brooks no interference with anyone's right to procreation, it does not preclude the imposition of reasonable "time, place, and manner" restrictions. The key decisional factors in terms of reasonableness are that the restrictions must be the minimum necessary to accomplish an important public objective, and, in no event, can they absolutely prevent the expression of a birth. In other words, reasonable "time, place, and manner" restrictions are ones that shift the time, place, or manner of the birth, but not the fact of the birth.

It is, for example, a reasonable "time, place, and manner" restriction on the second bioethic of birth to require that births be accomplished with banked sperm. This is reasonable because it does not deny any intended birth. Yet it prevents virtually all unintended births, which are conditions of disease, not of intentional expression. Another fair time, place, and manner restriction on the freedom of procreation is that all shipments of sperm be accompanied by objective information regarding services that test the sperm's propensity for genetic diseases and options concerning egg-cell testing or transgenic treatments. Again, this restriction is reasonable because it does not deny the freedom of any particular birth. Yet it accomplishes the important public health objective of "truth-in-birthing" for both parents.

Many of us are familiar with the need to get a permit for a parade and would justifiably shudder at the prospect of having to get a permit for a birth. The bioethics of birth would never countenance birth permits because they do not serve any legitimate purpose. Parade permits enable cities to clear the relevant streets for a march, organize police escorts, and broadcast traffic advisories. None of this is needed for bringing a child into the world. If a government alleged that it needed to know how many students would be coming to school in the future, it would get the numbers soon enough from the facts of actual births. If a government alleged that it needed to limit its population, it would be in violation of the second bioethic of birth by doing so via the suppression of births. Instead, inocuseeding will itself accomplish population policy goals by eliminating the 50 percent or more of births that are unintended.

It would even be wrong for governments to claim they needed to authorize births so that sperm was released only to married couples. This would be much more than a "time, place, and manner" restriction, because it would be an absolute prohibition of the birth rights of single people and gay couples. Just as brilliant speeches have come from jailhouse scholars, beautiful children have come from impoverished unwed mothers. The difference between a right and a privilege is that no value-based judgments are allowed to interfere with the exercise of a right. The freedom of birth is a right, not a privilege, and it is a right of every person regardless of marital status, sexual orientation, or socioeconomic status.

Now it may be argued that inocuseeding is an unprecedented expansion of government power and an unprecedented intrusion upon people's lives. Indeed, many will deny that the word "inoculate" can be used in conjunction with "vasectomy." But these claims are simply not valid and do not hold up to comparative analysis. We have been living under a male-dominated power structure for so long that we are afraid to call unintended pregnancy what it is—a preventable pandemic disease. Government programs as large as the one proposed here are routinely implemented for diseases far less widespread than unintended pregnancy. Those diseases, however, afflicted men as well as women, and were not camouflaged with religious gobbledygook as "God's will."

It seems hardly consistent to argue that God created man in his image, as the Bible claims, and then that, although God did so intentionally, mankind cannot do the same. Of course, this inconsistency is largely born of the practical impossibility of universal intentional childbirth prior to the biotechnology revolution of the 1990s. But advances in sperm banking, genetic testing, and in vitro fertilization have now made intentional childbirth always possible. Accordingly, we may now reproduce in God's image: intentionally. We need only to cleanse our religious doctrines of precepts based on old technology and to summon the corresponding political will to do the right thing. In so doing, men must overcome fear of losing that age-old indicium of male privilege: "I can make you pregnant but you can't make me."

As for inoculation, the concept has always meant that it is wiser to inflict a very modest harm on a person in order to avoid even the small possibility of a much larger harm. Hence, some inoculations produce minor skin irritations, and others may even make us a bit sick. But, still such inconveniences are considered a small price to pay to avoid even a small chance of coming down with a devastating disease such as polio or the greater chance of coming down with measles or mumps. It is exactly the same situation with sperm banking followed by inoculative contraceptive vaccinations or vasectomies. There is a very minor harm—the vaccination or vasectomy—with the corresponding avoidance of a much greater harm—literally millions of unintended pregnancies per year. The vasectomy is a very minor harm because, unlike in the past, it will not prevent anyone from producing offspring. Sperm is safely available at any time during a man's life from one of two redundant sperm banks. And, consistent with the bioethics of birth, the law will provide that the sperm be made available to the donor at any time, with no questions asked. The avoided unintentional pregnancy is a major harm avoided. Because it occurs to millions of women a year, it often devastates lives and it underlies numerous social problems. If one believes that the fetus is independent human life, then the case for inocuseeding becomes yet more compelling. Inocuseeding avoids actual death for millions of fetal lives by obviating the need for abortion. Sperm banking followed by vasectomy or contraceptive vaccination is truly an inoculative solution to the plague of unintended genomic disease.

By way of comparison with other similar government programs, at its peak in 1950, polio claimed about 30,000 American victims—almost 1 in every 5,000 persons or 1 in 1,000 children. Most of those afflicted recovered quickly with no lasting effects, but some were paralyzed for life. Based on this level of incidence, the government launched a mandatory program to vaccinate everyone against the various forms of polio. Happily, there is virtually no polio in America today. But even at its worst, the scourge of polio never claimed so high a toll of suffering, ruined lives, and even death as unintended pregnancy does today. According to the Alan Guttmacher Institute, a highly respected reproductive-health think tank, in the United States alone, each year there are *3 million unintended* pregnancies, afflicting about 1 out of every 50 women. As with polio and other diseases, most women recover quickly, but hundreds do die or become permanently disabled, and countless thousands have their life plans essentially ruined. The *Washington Post* reported on June 15, 1996 that a new UNICEF worldwide study founded 600,000 women die *each year* in mostly unintended pregnancies and childbirth, plus another *18 million* suffer injuries annually. None of the dozen or so diseases we are inoculated against exacts so severe a quantitative or economic toll as does unintended pregnancy.

Clearly, we do not generally recognize the harm of unintended pregnancy as on a par with other diseases. This is so because women are a political and economic underclass in every society, who have only in this century won the right to vote (and that still not everywhere), and women are still laboring under thousands of years of ideological brainwashing that their primary purpose in life is to produce sons. In addition, only in recent years has biotechnology provided society with the ability to control the procreation process practically. In other words, uncontrolled pregnancy has just now become a *treatable* disease because only now can science allow society control over the insemination process. The government can require us to be shot up with weakened pathogenic microbes from various animals to prevent unlikely but severe human suffering. Accordingly, there is ample precedent simply to require us to store our sperm safely and avoid the near certainty of the massive human suffering brought on by genomes we did not intend to create.

It is also interesting that an inocuseeding program does not involve such an unprecedented change in habits as might have been thought. The graph below (Figure 2) shows that as people get older they shift their contraception technique from

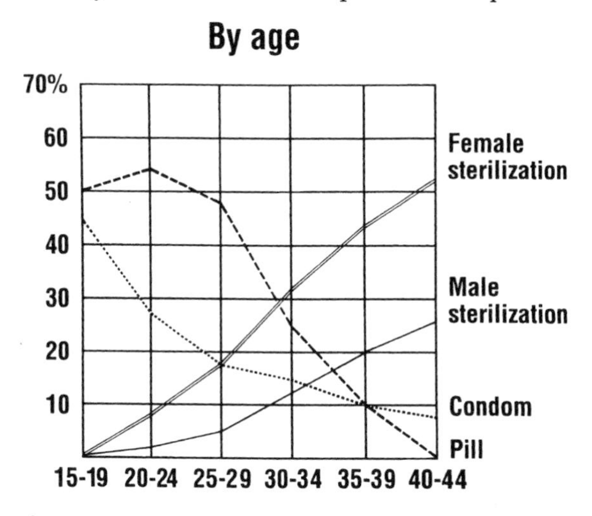

FIGURE 2: Participation in Pre-Inocuseeding Contraceptive Technologies.

Source: National Center for Health Statistics, 1990.

Note: Percentages not accounted for on the graph represent other methods of contraception, including diaphragms, IUDs, and periodic abstinence.

temporary to permanent methods. Unfortunately, the people using temporary methods are responsible for much of our epidemic of unintended genomes. Young people have naturally been unwilling to use permanent contraceptive methods since they have not yet had a chance to exercise their genomic expression. Inocuseeding will guarantee all young people of their genomic expression rights despite their use of a permanent contraceptive technique. Accordingly, all that inocuseeding requires in terms of a change of habits is roughly tripling the existing occurrence of vasectomies and time-shifting them from middle age to puberty. Both of these changes are enabled by the mandatory free sperm-banking feature of inocuseeding and by the development of contraceptive vaccinations.

Two final concerns may be why only men are inocuseeded, and why all men are inocuseeded. With regard to the first, it is simply a matter of medical efficacy and common sense. An inoculative vasectomy is a quick, noninvasive procedure with a zero fatality rate and a vanishingly small error rate. "Inoculative" reproductive techniques for women do not really exist. For example, unfertilized egg cells cannot be stored for a long period, making the female version of a vasectomy, called tubal ligation, not even inoculative—all pregnancies are prevented, intended as well as unintended. Also, tubal ligation is an invasive medical procedure with a much greater risk of harm than vasectomy. (Notwithstanding these facts, India now manages to sterilize about 4 million women per year under horrid conditions, and its population is still surging past the billion mark. It would be far more effective and ethical to inocuseed all adolescent Indian boys, an approach now made more problematic by India's history of forced male sterilization in the 1970s.)

Biochemical techniques, such as the pill or implanted devices, have a history of causing medical harm in significant numbers of people. Therefore, they fail the "minor harm" criterion for an inoculative approach. Insertional contraceptive devices, such as the IUD or condom, are either not sufficiently effective against unintended pregnancy in actual practice or they cause their own health problems. Developing and implementing a safe approach to inoculative reproductive contraception that could apply to both egg- and sperm-bearing people would be in society's interest in eliminating unintended genomes. Presently, however, sperm banking with inoculative vasectomies is the only medically efficacious approach available.

In all likelihood, as support for inocuseeding grows, our masturbation-squeamish neopatriarchal society will quickly come up with an alternative to postpubertal masturbation followed by vasectomies. In the spring of 1996, the National Academy of Science's Institute of Medicine reported the results of a lengthy contraceptive technology study involving government, industry, and insurers. In addition to confirming that current techniques are so problematic that "57 percent of pregnancies [in the U.S.]—more than three million each year—are unintended," the study noted that new developments in molecular biology could result in a vaccine-type approach that targeted "the specific genes recently found to be essential to contraception." Such a vaccine could be turned off in a petri dish when conception is desired, subject to preimplantation genetic diagnosis, or it could be turned off via an injection if contraception is desired via intercourse. The study also noted that such a contraceptive would be developed only if Congress immunized its developers from product liability lawsuits, a request the institute had already made back in 1990.

An alternative neonatal-based means of implementing inocuseeding may be even easier for society to swallow. This alternative arises from the 1996 discovery by Dr. Ralph Brinster of the University of Pennsylvania that sperm *stem* cells can be immortalized. Briefly, human sperm stem cells are present at birth and begin differentiating into active sperm cells in puberty, or on being placed in an appropriate laboratory solution, or even, remarkably enough, on being placed in the testes of another mature mammal, such as a mouse! Hence, at the time of neonatal circumcision, an inoculative vasectomy could be performed, with sperm stem cells banked until the individual is ready to have children. At that time, the sperm stem cells could be developed into active sperm, which could then be used to inseminate a woman directly or fertilize an egg cell in a petri dish. This approach has all the benefits of inocuseeding but avoids the squeamish issue (for a purportedly puritanical society) of postpubertal masturbation followed by teenage vasectomy.

Concerning the need to apply the third bioethic of birth to everyone, this is an absolutely essential criterion in order to avoid social eugenics. Were there any exceptions to universal applicability, the specter would exist that only certain groups of the population were being "sterilized" while leaving other subgroups of the population free to propagate. This would not be true because everyone's sperm would be available on demand. Nevertheless, the appearance given would be one of society taking steps to reduce the fertility of certain demographic groups (those inocuseeded) while not taking steps to reduce the fertility of others (the uninocuseeded). Such appearances would undermine public support for the entire disease-prevention program. In addition, any persons excluded from the inocuseeding program would, in fact, be disease vectors within the population at large. This means that they could readily inflict unintentional genome gestation on unsuspecting women.

Legal precedent also supports the requirement to apply universal inocuseeding to all men. In 1905, the United States Supreme Court in *Jacobson v. Massachusetts* affirmed the constitutionality of compulsory vaccination, subject to the understanding that it would apply to all the citizens of a state, not just to one or another ethnic group. Rejecting the claim of Henning Jacobson that vaccination against smallpox impinged on his constitutional rights, the Court ruled that in matters of public health the interests of the many would have to outweigh the idiosyncrasies of the few. The same case also disposed of another likely objection to inocuseeding: irrational fears. In *Jacobson v. Massachusetts,* Henning Jacobson feared he would become very ill from the inoculation. In our day and age—typified by President Clinton's 1994 firing of his Surgeon General for suggesting that schoolchildren be taught about masturbation—we can be certain that some people will claim that masturbation is a major harm (or even a sin) in itself. The Supreme Court determined that irrational fears were not a defense against a mandatory inoculation program, although the possibility of exceptions was established for valid health concerns and religious objections. Similar exceptions might carefully be carved out of a universal inocuseeding program. Even programs that are 90 percent effective will be well worth the effort in terms of freeing women from genomic disease and drastically reducing the incidence of abortion.

Notwithstanding the legal precedent in support of inoculation, it might be argued that coital reproduction enjoys a special legal status that cannot be impugned. For example, procreative-liberty advocate John Robertson, in his 1994 treatise *Children of Choice,* argues:

> Married couples would have a fundamental constitutional right against state limits on coital reproduction, whether it takes the form of penalizing them for having more than a set number of children, requiring licenses to parent, or mandating sterilization, contraception, or abortion. Restrictions on marital reproduction are theoretically possible only if the state can show great harm to others from the reproduction in question. A situation that might justify such a limitation would be overpopulation, but such a restriction would have to be equitably distributed and structured to minimize coercion and unwanted bodily intrusion.

As is often the case from nonfeminist perspectives, Robertson fails even to consider the huge toll of unwanted pregnancy as a "great harm to others." As we have seen, the harm from this epidemic far exceeds that of other diseases for which vaccination or inoculation is well accepted. Furthermore, unlike Robertson's Orwellian scenario, the implementation of inocuseeding imposes no limits on reproduction whatsoever, only on *coital* reproduction. Hence, inocuseeding should be justified even under a procreative-liberty analysis, because the harm avoided is great (the epidemic of unintended pregnancy) and the interference with reproductive rights is minimal (just time, place, and manner). Indeed, Robertson concludes that, for unmarried couples, the constitutional "right to avoid pregnancy and reproduction [via contraceptives] . . . does not necessarily imply a right to engage in coitus in order to get pregnant." In short, there is no constitutional right to make somebody pregnant.

There is no doubt that, in the 1990s, universal inocuseeding sounds like a radical solution to the problem of unintended pregnancy. Nevertheless, on closer examination the radicalism fades away. Inocuseeding sounds radical mostly because a patriarchal ideology of thousands of years' duration has brainwashed us into thinking of pregnancy as woman's duty, job, or curse. So unintended pregnancy only looks like a disease after we remove our patriarchal "blinders." This is hard to do since those "blinders" are pretty much welded to our minds. As many a wit has observed, pregnancy would have been "cured" long ago if it afflicted men as well as women.

A secondary reason the third bioethic of birth seems radical is that we are not accustomed to the notion of involving third parties (sperm banks or fertility centers) in the process of birthing children. This, however, merely reflects the newness of genomic technology and a lack of historical perspective. Prenatal genomic analysis technology is less than twenty years old and still in its infancy. As we are gradually able to influence our offspring's health positively before birth, future generations will consider us "primitive" for having birthed children without the benefit of pregestation genomic analysis and modification. Just as it is "wrong" to let our children go to school uninoculated against some diseases, it will be "wrong" to let them finish school uninoculated against the disease of unintended genomics. Just as it is in our best interest to get our various shots, it will be seen in the future to be in our best interest to bank our sperm and genetically "tune" ourselves prior to conception.

As to historical perspective, in less than a century we have gone from virtually no physician-attended births to doctors' nearly always being present. It took only the fifty years from 1910 to 1960 for childbirth hospitalization to go from rarity to normality. And today, about 70 percent of U.S. pregnancies undergo routine prenatal screening. Twenty-five years ago this technology was all but unknown. We have very rapidly become accustomed to involving third parties in the handling of our seedlings of sex. Indeed, according to Congress's Office of Technology Assessment, upward of forty thousand children *each year* are born using the very inocuseeding techniques described in this chapter. Clearly, there is no evidence that humans are intrinsically resistant to the "third party" aspects of universal inocuseeding.

treating disease

Notwithstanding society's best efforts to ensure that all births are intentional, consistent with our highest bioethical standards, there will clearly still be unintended pregnancies for many years to come. Most of these pregnancies will be due to uninocuseeded men, and a few will result from unsuccessful contraceptive vaccinations or vasectomies. However, we must also consider the likely growing numbers of embryos—either test-tube or implanted—that are unintended because of their genetic readout.

Suppose a person intentionally conceives life, but later finds out that the fetus is badly deformed. The mother now wants an abortion. Is this the infliction of death on a citizen-to-be or the curing of a state of disease in a citizen-at-hand? The logically consistent answer is to comply with the mother's wish for an abortion. So long as pregnancy is either health or disease, based on the mother's state of mind, it should be clear that this state of affairs can change based on new information. In essence, the mother is saying, yes, she intended to conceive life, but not that particular form of life. Therefore, she is suffering from disease, and society owes her the minimal health care of a no-questions-asked abortion.

Critics will argue that this approach places the whims of the mother above the life of her child. Most experts agree, however, that it is not possible to set a point at which the fetus becomes a "child" as opposed to an outgrowth of maternal tissue. After exhaustively reviewing the literature, fetal policy expert Steven Maynard-Moody declared in his 1995 treatise, *The Dilemma of the Fetus:*

> Defining the beginning of human life or personhood at either birth or at conception benefits from simplicity, being based on specific and easily recognized events. But both views ignore the complexity and significance of fetal development. To find a milestone, a point, during the stages of fetal development when the fetus becomes an alive person may not be possible.

In other words, for at least part if not most of a pregnancy, the fetus cannot be considered an alive person—it lacks survivability, brain waves, or other criteria we use today to determine whether a postbirth person is alive or dead. Clearly, for that period, if a mother views the tissue growing within her as alien and unwanted, it is a disease and excision via abortion is the cure. While we are not used to calling unwanted pregnancy "disease," classic work such as George Rosen's 1958 *History of Public Health* make clear that what society considers to be the causes and definitions of disease have fluctuated many times throughout history—usually in response to conceptual or technological breakthroughs.

Once the fetus is viable, in the third trimester, it is a rare mother who wants the fetus removed. In the vast majority of cases, such a decision would only be made for the safety of her own life, or because prenatal scans have indicated the fetus has a major health problem, or because she never had the option of an abortion earlier in an unwanted pregnancy. For many a mother the fetal health problem may be so inconsistent with her procreative intentions that she would rather abort. In such a case, the medical profession should be obligated to cure her of her unintended pregnancy even at the expense of a lost fetal life. To decide otherwise is to convert a definite life, the mother's, into a vessel for a problematic life, the fetus's. This would be wholly inconsistent with the bioethics of birth, because it is the parent, not the offspring, whose procreative expression deserves absolute protection. If it is safer to birth the fetus, then, of course that is the ethically advisable course. We can mourn a lost viable fetus as an innocent victim in the war against disease, with the enemy being our own society's failure to: (1) achieve universal inocuseeding, (2) guarantee meaningful genomic health care services to all, and (3) teach appreciation for the full diversity of human genomic expression. Anger directed at an abortion clinic is as misdirected as anger at a MASH doctor triaging wounded victims of an enemy artillery attack.

Society can minimize the occurrence of intentional pregnancies evolving into diseased pregnancies by making preimplantation fertilization and genetic testing services widely available. Already, a January 1994 Time/CNN has poll reported that 58 percent of adults (with a + or −4.5 percent sample error) want their unborns tested for "genetic defects." Genetic testing of both the sperm and egg donor will provide a fairly good picture of what the likely genomic consequences of fertilization will be. Of course, it is up to each parent to decide the scope of their genomic intentions. Many parents will be happy with any genome that develops. Some do not intend to conceive genomes plagued with serious, debilitating conditions. Some will be so particular that they intend only genomes with certain superficial demographic characteristics. It's all a matter of freedom of procreation: the second bioethic of birth.

Once the child is born, the disease of pregnancy is no longer curable by eliminating the fetus. There no longer is a fetus, there is a living, breathing person. If the mother does not want the child, then adoption should be a readily available alternative. But adoption is not the answer to an unwanted pregnancy. Adoption does not cure the disease, it only manages an unavoidable consequence of the disease. Abortion is the only treatment for the disease of unwanted pregnancy. The main cure is prevention, namely, universal inocuseeding. A secondary cure is prefertilization genetic analysis to avoid creating embryos that a parent may not intend to keep.

Finally, there remains the issue of the bioethics of birth as it relates to embryos formed in a laboratory. Can they be bioethically terminated? Can they be bioethically formed? If so, when and for what reasons? The creation of an embryo is the creation of a new version of the human genome. Pursuant to the first bioethic of birth, the human genome is the common heritage of humanity. This means that everyone is free to use the human genome for peaceful, beneficient, nonpollutive purposes. However, is the creation of several embryos so that one can "pick the best" beneficient? Is the creation of embryos for reasons other than intentional birth nonpollutive? Is the creation of embryos for parentless laboratory birth to a corporate entity a peaceful purpose consistent with the common interest of humanity? These are the boundary questions that we next address.

It is possible that, a laboratory may be able to artificially gestate an embryo fertilized with anonymously banked sperm and egg cells to the point of life. Recent experiments have shown that the fetus itself manufactures what is necessary to create its own womb, and that even male baboons have been made pregnant. Accordingly, with technology such as heart–lung machines and artificial kidneys, laboratory gestation of a fertilized embryo is not at all far-fetched. The laboratory may intend to sell parts of the fetus the way transgenic organs will soon be sold. The laboratory may even want intentionally to conceive new life. After all, even corporations have the right to freedom of expression. Why not also the right to freedom of procreation? And, to highlight the dilemmas yet further, the genomic fertilization laboratory need not be some aseptic subsidiary of a Fortune 500 corporation. It may be a community-based, religion-based, or even a family-based affair. Should these extrapolations of extrabodily biotechnology be allowed?

Once again, we must refer to the bioethics of birth for guidance. The first bioethic tells us that the human genome is our common heritage. It may be used but not abused. Creation of genomes for spare parts sounds like abuse. It sounds suspiciously analogous to melting the polar ice caps to produce more ocean or depleting the atmosphere to create more space.

These things are simply not allowed. Creation of genomes for parentless children also sounds like abuse. Mass production of unloved genomes does no honor to our past. It just pollutes our future.

The second bioethic of birth tells us that each person has an inalienable right to express him- or herself through the human genome. But the creation of genomes for spare parts does not seem like meaningful expression. The First Amendment does not protect the right of someone to set up a loudspeaker and blast out the letters of the alphabet all day. Neither would the second bioethic of birth enable genomes to be created for tissues, hearts or lungs. The creation of genomes for parentless children would also appear to be a mindless application of the second bioethic of birth. The First Amendment is for heartfelt messages, whatever they are. The second bioethic of birth is for heartfelt children, whatever they are. A reasonable "time, place, and manner" restriction for the government to place on artificial gestation is that there be at least one person who is contractually committed to loving the life she or he causes to be created.

The third bioethic of birth enables society to take reasonable steps to stamp out genomic disease before it starts, without infringing upon each person's right to genomic expression. Because the creation of embryos other than for intended life is not a protected genomic expression, the government can readily quash this practice in full consistency with the bioethics of birth. As for the gestation of embryos outside a human body, if some persons want to do this, it should be their right. If a corporation or other group wants to do this, the government can reasonably close down any operation that looks more like genomic pollution than parental genomic expression. After all, analogies take one only so far. The production of life is ever so much more weighty than the production of words. Although any corporate entity is entitled to the production of words, we can reasonably allow only individual parents to be entitled to the production of life.

Finally there is the issue of a person creating several embryos so that only the "nondiseased" one can be selected. For example, most crippling diseases require *two* particular genes to be present in the same "garbled" form in order for the disease to actually affect the body. Suppose an egg donor and a sperm donor are both found to carry one of the crippling disease genes in their genome. This means that the donors don't actually have the disease (because two genes are needed for it to be expressed); the donors are just "carriers" of the disease trait.

Now, recall that each egg cell and each sperm cell has half a genome. It is a completely random, 50–50 chance as to which particular half of the genome any particular egg or sperm cell has. Accordingly, this couple would have a 1 in 4 chance of producing an embryo certain to come down with the crippling disease, and a 1 in 4 chance of producing an embryo that was not even a carrier (i.e., the embryo took either the diseased gene from each parent, or the nondiseased gene from each parent). There would also be a 2 in 4 chance of producing an embryo that, like its parents, was a carrier but would not actually get the disease (i.e., the embryo took a disease gene from the egg and a regular gene from the sperm, or vice versa).

Based on the foregoing scenario, there are five possible reproductive options, only one of which involves laboratory fertilization of multiple embryos. The other four options are: (1) create just one genome and accept the baby however it is formed, certain to get the disease, mere carrier of the disease trait, or otherwise, (2) create just one genome and abort the fetus if prenatal testing indicates it reflects an unintended 1 in 4 diseased genetic configuration; (3) create the genome using someone else's sperm and egg that do not carry the disease trait, thereby eliminating the risk of disease; or (4) create the genome as a clone of one of the parent's non-germ cells, in which case it will be just a carrier of the disease trait and will look like the cloned parent used to look. We have analyzed each of these four options elsewhere in this book and found them to be consistent with the bioethics of birth. If parents select none of these four options, it is because their personal eugenic drive is too strong to risk a diseased baby or to use half of someone else's genome, or because they have a personal moral objection to cloning or to aborting a fully gestating fetus.

The fifth option available under the scenario described above is to fertilize several embryos in the lab and select for implantation only the one without disease carrying genes. Since there is only a 1 in 4 *random* chance of this occurring, we would expect to see this configuration if four embryos were created, but it could occur with fewer and it might not occur even if many were created. Also, since simply carrying one of the two genes necessary for disease-bearing does not make the embryo or child in any way less healthy, the parents may be completely satisfied with the 2 in 4 random chance of an embryo forming with just one disease-carrying gene.

The new bioethical issue to be addressed is whether the nonselected embryos are a "disease" to be terminated or to be kept frozen for later consideration. The third bioethic of birth tells us that genomes may not be formed for purposes other than intentional childbirth. Accordingly, the parents may bioethically implant only one out of several embryos if that is their intent, but what becomes of the other embryos? Because the third bioethic of birth proscribes unintended genomes, a condition of creating more than one embryo outside the body must be that specific parental intent is given for each. The two possible intents are implant one embryo and terminate the others, or implant one embryo and freeze the others with the intent to implant one or more later. Bioethical questions arise with regard to intent to gestate embryos at a later date because that date may be left unspecified (never come) or, if specified, may occur in a time frame in which the parents could not nurture and care for the child (such as after the parents' death).

The second bioethic of birth does not protect the right to create embryos that cannot be nurtured and cared for by a parent. The second bioethic of birth is the right to a person's *parental* expression of will, not just the right to create genomes per se. By analogy, my freedom of speech is *my* freedom to speak, not my freedom to make you speak for me. In other words, we cannot bioethically create embryos in limbo. There must either be a specific contractual commitment to nurture and care for the embryo within a few years, or it should be promptly terminated. The precise number of years of intended embryo storage should be specified by the parent(s), up to a legal maximum, such as ten years, as a precondition to any biotechnology laboratory's actually creating an embryo. Embryos should not be permitted to outlive the people that intended to gestate them. That would no longer be intentional life.

We have seen here that the corollary of our freedom of procreation is society's obligation to minimize mistakes by preventing, curing and controlling unintended life. (Whether or not life is intended—that's up to each individual childbearer.) Government can and should rightly make this so through universal inocuseeding, readily available abortions for the unintended results of sex with the uninocuseeded, and liberally available genetic counseling and fertility services. The government's obligation with regard to making genomic services available applies with special vigor to those out of the economic mainstream. Class differences, left unaddressed, will create de facto social eugenic outcomes in the era of reproductive technology. Accordingly, the government has a social obligation to ensure a parity of genomic technical capability to all people, without regard to their economic status. With this third bioethic of birth, including government assistance to the economic underclass, abortion will be very rare because sex won't cause babies and birth anomalies will either be remediable before gestation, or accepted as beautiful variations on the theme of life.

a note on unintended life after birth

Nearly everyone alive today is pretty much unintended life. Whether or not our parents wanted a child, in most cases that wasn't the main thing on their minds when they went to bed the night we were conceived. Hence, the only stigma about being an unintended child is the stigma some parents unfortunately impose on their children. Prior to universal inocuseeding, virtually all children have been unintended kids. We love them just the same, completely without regard to whether or not they were intended. As discussed in Chapter 7, the humanity of all persons attaches to them upon their birth, without regard to how they arrived.

Once a person is born, that person is no longer part of the mother's body, and she or he is therefore no longer in any conceivable way "disease." In the overwhelming majority of cases, we love the child we birth, regardless of whether or not we intended to have one. Once the child is born, there is a new person that carries much of our body and is ready to be imprinted with all of our soul. The birth of a child is a wholly separate event from the reality of unintended pregnancy as a condition of disease. Our subsequent love for a child in no way diminishes our right to be cured of disease. Were it otherwise, we would spend our entire lives gestating one child to be loved after another. In other words, children are never disease.

chapter 9 respecting our genome: ending the crime

According to most estimates, everyone carries at least five to 10 genes that could make him or her sick under the wrong circumstances or could adversely affect children.

Scientific American, June 1994

We're all mutants. Everybody is genetically defective.

MICHAEL M. KABACK
University of California geneticist, *1994*

The fourth bioethic of birth makes it explicitly clear that governmental discrimination regarding the genetics of our offspring, either directly or indirectly, positively or negatively, amounts to a crime against demography—democide. More subtle but no less pernicious than genocide, this new crime of democide is not only a prohibition from interfering with our procreative freedom—it is a call for universal action to help beat back any effort, anywhere, that has the effect of diminishing the demographic diversity that blesses our species.

The holocaust of sex going on in Asia today, as described in

Chapter 1, is an example of democide left uncontrolled, of our genome left naked and without a proper defense (except for eventual "marketplace demand" for females). It is no anwer to say that in China sex-selection technology is illegal, or that in India the disappearing women are the victims of choice. The fact of the matter is that in each instance the government has taken measures that make the holocaust of sex wholly predictable, if not inevitable. The slaughter of Jews by Ukrainian peasants was no less the fault of the Nazis simply because they didn't fire the guns. The genocide of the Native American was no less real if due to disease, cold, and hunger rather than a congressional order.

Possible holocausts of the gay, the obese, and the cancer-prone—incipient in surveys that show 10 percent or more of the population would abort such genomes—must be stopped before they start. These holocausts would be the result of government-sanctioned discrimination, and they can be bioethically prevented only by the government's prohibition of any such discrimination. One cannot blame parents for wanting only the best for their children. If they live in a society in which the government sanctions laws denying gay people the right to marry solely because of their sexual orientation, why bear gay offspring? If they live in a society in which the government permits the obese to be denied a job solely because of their size, why continue the pregnancy of a child likely to become obese? And if they live in a society in which the government allows health insurers to deny coverage to persons prone to cancer, why bring such suffering and humiliation into the world? Prejudice and discrimination are the proximate causes of democide as well as genocide.

Bioethicist Susan Wolf, writing in the Winter 1995 issue of the *Journal of Law, Medicine and Ethics,* offered a fresh and vivid new perspective on how best to remedy the kind of discrimination that leads to democide. She makes clear that, like racism and sexism, the chief problem is one of getting society to see people as people, not as skin tones, genitals, or genes:

> Combating individual instances in which a person of color was treated differently because of race, or a woman treated differently because of her sex, will not uproot deep and widespread racism and sexism. Similarly, condemning individual instances of genetic discrimination will do little to address systematic genetic categorization, a world view that seeks to sort people by their genetics, and the conviction that supposedly deviant genes merit different treatment.

Building on Wolf's thesis, we can see that the problem is pervasive "genism," an analog to racism and sexism. Each of these malodorous ideologies fails to embrace the continuum of humanity, the fact that we are all part of one diverse whole. Instead, the "genist" sees people as falling into certain categories based on their genes, some better and some worse, and believes it to be acceptable to treat different categories differentially. In Wolf's well-reasoned view, nondiscrimination policies are not enough to prevent democide. To the contrary, genetic differences will be used to justify democidal policies.

The fourth bioethic of birth requires government to be proactive in fighting genism. It requires policies that make palpable for all of us, especially children, that we are all part of the human genome. That an attack on part of the genome is an attack on all of us. We must come to understand that solidarity with all genomic variants is our solemn obligation, and that genism is as wrong and self-destructive for society as are racism and sexism.

The obligation of social solidarity beyond genetics applies not only to naturally occurring variations but to transgenic creations as well, that is, to persons with some nonhuman genes. In a prescient article in the June 1992 *UCLA Law Review,* "A Theory of Constitutional Personhood for Transgenic Humanoid Species," Michael Rivard observed:

> In the past, African-Americans were legally enslaved and denied any constitutional rights on the basis of skin color. Congress (eventually) recognized that race is not a meaningful characteristic by which to differentiate constitutional persons from nonpersons. If one agrees that it is unacceptable to deny constitutional personhood to other humans solely on the basis of race, then one should also agree that it is unacceptable to deny constitutional personhood to other species solely on the basis of genetic compostition.

Rivard concludes that so long as a being is aware of and appreciates legal rights, it is entitled to those rights. Accordingly, transgenic humans should have the same human rights as other humans. Their personhood is unaffected by their genetics.

Demographic death is a crime against humanity no matter how it occurs. Accordingly, the fourth bioethic of birth demands that we all learn from childhood that democide, like homicide, is wrong. The authorities are responsible for demographic death whenever genomes start to disappear. Democide will not occur if the government's clear message, backed up with the force of law, is to respect people as individuals, not as members of some arbitrary sociogenetic group. Democide, like genocide, is simply discrimination and bigotry run wild.

stopping the holocaust of sex

Researchers looking into the holocaust of sex have uncovered several causal factors: stereotyped notions of female inferiority, sexist governmental policies and social customs, local manufacture of ultrasound machines, widespread availability of amniocentesis, liberal abortion policies, government family-size policies, and corrupt medical practitioners. The millions of missing women cannot be brought back, but ongoing democide can be ended. The main cause of the holocaust of sex is the low status of women. This sex-based caste system is the result of bigoted views that someone able to gestate is mentally different from someone who cannot. In short, the low status of women is due to an officially sanctioned apartheid of sex—a separation of people into two legal categories (one dominant, one repressed) based upon their genomic abilities. Curing this social bigotry takes two major steps:

1. *Abolishing all social, legal, and economic distinctions based on the genomic identity of a person.* In particular, people able to gestate must have the same rights to work, marry, and wield power that people unable to gestate have. Job, marriage, and other applications should not ask, and should not care, about sex. The X or Y's of a person's genome must be as legally insignificant as the other 22 chromosomal gene sequences.
2. *Taking action to cure the biggest disease that affects women disproportionately: unwanted pregnancy.* It causes no harm to men to implement universal inocuseeding, but potentially saves nearly every woman from one or more bouts with a life-threatening, career-derailing,

> mentally traumatic disease. Conception must always be a choice, not a happenstance.

The principal factor causing female democide in China, India, and elsewhere in Asia is the widespread discrimination women face. Before ultrasound machines mechanized the holocaust, female infanticide was practiced widely. Prior to the Communist Revolution, Chinese women frequently were not even given names—they were, instead, the "sister of A" or the "daughter of B." Until today, the widely practiced dowry custom in India required families to pay huge sums of money to the man who married their daughter. This money was lost to the family because, unlike money spent on education, it does not benefit the daughter.

Men overwhelmingly control the wealth, the power, and the educational opportunities in Asia. Women are considered lost to the family upon marriage—all that was spent on their development is a waste. In India, there is the saying that having a daughter is simply watering a neighbor's plant. Millions of people logically decide that it is better not to be born than to be born a woman. On the other hand, a single female child is often still desired. This situation is summarized in an exhaustive 1994 *International Family Planning Perspectives* article by T. Rajaretnam, deputy director of India's Population Research Centre:

> Most couples desire at least one son, often two, and want either one daughter or have no explicit preference for daughters at all. The stronger preference for sons over daughters is because sons are thought of as productive assets for labor, as providers of security in emergencies and in the parents' old age, and as conduits to carry on

> the family name and perform family rituals. On the other hand, daughters are thought to be a liability, as their marriages cost the parents a large dowry and they move to their husband's home after marriage. The preference for one daughter is presumably based mainly on emotional grounds, as well as on the expectations that she will attend to household chores before marriage.

Evidence to back up this hypothesis was found after years of field research. For example, women would refuse to use contraceptives until they had achieved a desired sex ratio. If luck ran against them, and they became pregnant with a second daughter, the best alternative was democide. The researchers, Rajaretnam reported, found that the democide would stop only after one or two sons were born:

> Before accepting family planning, many Indian women try to ensure that they have had not only two or more male children, but also at least one female child. However, the desire for a female child appears to be not as strong as the desire for a male child, as is evident from the fact that prevalence [of contraception] among currently married women with no female living children is sometimes many times higher than the rate among currently married women with no male living children.

A similar situation was found in China. Beijing University's Zeng Yi reported in the June 1993 issue of *Population and Development Review* the following statistics of male-to-female birth ratios: Among firstborn children, 105 boys are born for each 100 girls; among second children, 121 boys for each 100 girls; among third children, 125 boys for each 100 girls; and among fourth children, 132 boys for each 100 girls. Clearly as parents produce more children, they become ever more desperate for boys. Had the boys they wanted occurred in earlier order births, they probably would have stopped having children. Having failed to deliver boys, the parents keep conceiving children and become ever more likely to engage in democide as their number of girls increases. With the "one family–one child" policy in effect, the ratio of all male-to-female births ranges from around 115 to 135 boys for each 100 girls, depending on the locality. With ultrasound machines in virtually every Chinese town, there is no doubt that democidal practices account for the millions of missing Chinese girls.

The solution to ending Asia's holocaust of sex lies in the major factor uncovered by researchers: rampant discrimination against females. For example, women must be thought of as equally productive assets for labor, as equally good providers of security in emergencies and in the parents' old age, and as equally valid conduits to carry on the family name and perform family rituals. Such changes are unlikely to occur so long as women can be made pregnant at the will of a man. Behind sex discrimination is the age-old susceptibility of women to the disease of unintended pregnancy. But biotechnology allows us to eliminate that disease. Accordingly, one key to ending the holocaust of sex is the third bioethic of birth—the imposition of universal inocuseeding as a public health measure against the disease of unintended childbirth. Although developing countries such as India and China often lack the public health infrastructure to support functions such as sperm banks, they also have the greatest motivation to allocate the necessary budget for inocuseeding. Indeed, population and development pressures have already catapulted China and India to the forefront of nations implementing expensive, government-directed anti-fertility policies—sterilization in India and one-child families in China. Unfortunately, these policies are both unethical and wrong-headed.

Government-imposed family size limits only add fuel to the holocaust of sex. In so doing, they also violate the second bioethic of birth. With universal sperm banking there is no need for government-imposed family size limits. Women who can choose when to get pregnant, and with what to get pregnant, are unlikely to get pregnant very often. However, if one can birth only one child, and if the sex of that child is important, then a "one family–one child" birth policy is a prescription for mass demographic death. The bachelor villages and overwhelmingly male classrooms of China are quite proof enough of the results of that policy.

Although some urge the outlaw of sex selection technology, this approach is morally wrong. It interferes with the right of people to express their genome as they please. It does not stop sex selection, but merely drives it underground. Outlawing sex-selection technology is a poor palliative for the holocaust of sex. It treats a symptom and ignores the disease. The illnesses are unwanted pregnancies and discrimination, not unwanted children and choice. Universal inocuseeding cures unwanted pregnancy and discrimination, while preserving freedom of procreative choice.

Historically, the male ability to set the agenda for pregnancy, and to make women carry out that agenda, created the vast superstructure of sexual discrimination that plagues women today. This corrupt socioeconomic artifice has enabled an insidious sexual holocaust to occur, one which not only bedevils women, but which threatens everyone on earth as the precedent for a long string of democidal wars. The same biotechnology that fuels demographic death could end it. We simply must summon the political will, and the plain common sense, to demand universal inocuseeding.

As mandatory inocuseeding takes hold, the socioeconomic and legal significance of gestation will evaporate. "No one can make me pregnant—I make me pregnant." People with vaginas will be every bit as able to undertake all of the economic, security, and social welfare tasks for their parents as will people with penises. The X versus Y features of genomes will not lose reality, but will lose criticality. It will be interesting to have XX offspring who can gestate, but they may choose not to. On the other hand, XY offspring may just use a surrogate to gestate, or may even gestate themselves, with delivery via cesarean section, as new research has shown to be possible. Genomic differences are a pleasant reality; they need not be an oppressive criticality. We can erase much of the basis for democide. We can stop fueling the holocaust of sex.

As Monique Wittig observed in her classic 1981 essay, "One Is Not Born a Woman," "Heterosexuality is a social system that produced the doctrine of differences between the sexes so that women could be dominated by men." There is much truth to what she says, for it is certainly possible that women can, in theory, refuse to be unwillingly impregnated. Instead, a mass psychology of sexual difference was developed, an apartheid of sex, that led women and men to think of themselves as intrinsically different. This mass psychology, which Wittig calls "heterosexuality," taught women to think of themselves as gestating beings while men thought of themselves as impregnating beings. From this simple social cleavage, comes the low status of women today.

The fourth bioethic of birth mandates government be proactive in eliminating social, economic, and legal practices that maintain genomic discrimination. With regard to the holocaust of sex, this means that people's birth certificates should not mention sex, that marital and family rights cannot depend on sex, and that no aspect of the socioeconomic infrastructure should be different for people with XY genomes from that for people with XX genomes. Coupled with the elimination of involuntary childbearing through universal inocuseeding—which is a medical practice—the holocaust of sex will be brought to an end.

Some may find frightening a world in which people with different genitals are found in the same washroom, or in which people with similar genitals are married to each other, or in which teenagers masturbate into a cup for subsequent intentional childbirth. However, there is nothing to fear here—nobody dies, and no peoples die, from intersexual lavatories or cogenital marriages or masturbation. What we should find frightening is a world in which genomic configuration is so important that millions of people decide it is better for *their own* genomic configuration never to live at all. That's the world we have today. It will only get worse—other people will try to wipe out different genomes—unless the fourth bioethic of birth takes hold.

preventing democidal discrimination

In early 1994, the National Academy of Science's Institute of Medicine, America's loftiest medical body, issued a comprehensive report on the implications of genetic testing for health and social policy. The report recounted a vast and growing array of genetic testing being undertaken, and concluded:

> There is a growing danger that the results of genetic testing, obtained by an individual primarily for purposes of disease management or for reproductive options, may be used by an insurer or employer to deny health or life insurance or employment."

Alarmed that many employers were jumping the gun and initiating genetic screening on their own initiative as a condition of employment, the Council of Ethical and Judicial Affairs of the American Medical Association has recently taken the position flatly that it is "inappropriate for employers to perform genetic tests to exclude workers from jobs."

Thanks to the medical community's warnings, governments are beginning to hear the message. In early 1995, the United States government decided to consider some limited kinds of discrimination based on a person's genome to be prohibited under the Americans with Disabilities Act of 1991. But much more broad-based protection will be needed. We need to legislate straightforward civil rights protection for the genome rather than adopt partial protection via backdoor disability law loopholes. As genetic testing is used as a basis for excluding workers from jobs, children from health insurance, and parents from life insurance, we can hardly be surprised to see growing numbers of people aborting embryos that test positively for any or all of the "unpopular" traits. In other words, we are rapidly sowing the seeds of the same kind of genetic democide here in the United States that is flaming out of control in Asia. There, it is a holocaust of sex. Here, it may be holocausts of the obese, the gay, the mentally different, the lacking in this, that, or the other chromosomal marker thought to indicate intelligence, loyalty, or whatever.

It is tempting to permit selective fertilization only for medical characteristics, to simply outlaw dickering with the seeds of sex for anything not considered an illness. Such an approach would not be right. It would violate the second bioethic of birth precisely because of the impossibility of separating medical from demographic characteristics. The second bioethic of birth is absolute—no one may interfere with our genomic freedom.

The proper solution lies in the fourth bioethic of birth—the duty of the government to use its best efforts to eliminate genomically discriminatory conditions. There is no reason for an employer to test a person's genome other than as a touchstone for discrimination. Even for an employer who claims to need genetic test information because of the public safety aspect of the job, such as for an airline pilot or heavy-equipment operator, the American Medical Association's Ethical and Judicial Affairs Council observes:

> Genetic tests are not only generally inaccurate when used for public safety purposes, but also unnecessary. A more effective approach to protecting the public's safety would be routine testing of a worker's actual capacity to function in a job that is safety-sensitive.

Genomic discrimination is wrong because it categorizes us on the basis of biology over which we have no control and which is irrelevant for any social, economic, or legal purpose. If we have a genomic predisposition to a particular condition that interferes with our job ability, then it is wrong to deny us that job until the condition manifests itself. To do otherwise is simply to help set the stage for democidal holocausts of persons with that genomic condition. This is exactly what occurred with women. It is what will occur with other sorts of people if we allow social eugenics to creep in through the back door of government-sanctioned genomic discrimination.

Some economists argue that genomic testing is needed to provide fair pricing of health and life insurance services. They argue that, since people do get ill or die at differential rates based on their genomic predispositions, then it is unfair for people with "healthy" genomes to subsidize the insurance of people with "unhealthy" genomes. There are three fatal flaws to this argument.

First, health and life insurance are social goods, like national defense, to which everyone is entitled in fair measure, regardless of whether or not they "deserve" it. National defense protects us from the enemy without. Health and life insurance protect us from the enemy within. In essence, this point follows from the first bioethic of birth, that the human genome is the common heritage of all humanity. Just because some parts of the ocean are more sensitive to environmental harm than others, does not mean we protect them any the less. There is a common interest in safeguarding common heritage resources, be they oceanic or genomic.

Second, it is in the interest of us all to encourage the full diversity of human genomic expression. We never know when that uninsurable or unemployable genome may be our own, or the product of our own. This point follows from the second bioethic of birth: Freedom of genomic expression is a fundamental human right. We permit all kinds of speech, even those with which we violently disagree, because we want to ensure that our form of speech is equally protected. Similarly, we must take care of all forms of genomes, even those that now seem much weaker than our own, because we want to make sure that our genome, too—no matter what the future holds—is also capable of being protected.

The third reason not to discriminate in basic health or life insurance premiums based on a person's genome is that doing so leads to democide, in violation of the bioethics of birth. For exactly the same reason, insurers are not allowed to charge African Americans higher insurance premiums than Euro-Americans, although statistically the former die at substantially higher rates than the latter. Were insurers to discriminate racially, they would further aggravate differential death rates by yet further restricting the access of African Americans to decent health care. The ultimate result is a democidal cycle of misery and death. The same situation prevails in genomics. If insurers discriminate genomically, then some genomic configurations will die at yet greater rates as a result of economically restricted access to health care, or the suffering of offspring left without family life insurance. The result is a democidal cycle of suffering, which finally plunges into extinction when parents decide it is better to abort the "unpopular" genomes. The fourth bioethic of birth is intended to prevent exactly this situation. The role of government is to eliminate discrimination so that people can freely choose to create genomic diversity.

Economics is a tool, not a decision. Economics can correctly tell us that the health care costs of some genomes will be greater than others. But economics does not tell us to make every genome pay its cost. That decision is an ethical one, and in the age of biotechnology it would be a serious mistake to start allocating socioeconomic rights based on a person's genetic blueprint. These kinds of decisions could start a chain reaction of democidal dominoes that would leave the planet a very ugly place, indeed. Our common genomic heritage, our respect for genomic freedom, and our condemnation of genomic discrimination must combine to outlaw any kind of differential economic rights to insurance based on the way we were born.

The bioethics of birth are a road map. They help us get through the vast moral desert of biotechnology and into the promised land of health and diversity. We need not accept the genie of death in order to work with the genie of life. We need not accept democide as the price of a better life. We can avoid the pitfalls of social eugenics, of sexual holocausts, and of demographic deaths. We need only summon the willpower to follow the right course.

The four bioethics of birth are companions. It's not easy to follow four scores, but we will not succeed if we do any less. The forces we are dealing with—biology, technology, society, demography—are too vast and too unwieldy to be managed with a simple "golden rule." A bioethical framework is needed.

First, there is the overarching recognition of the human genome as the common heritage of all humanity. It is not to be owned, polluted, narrowed, or destroyed, no matter the vision and no matter the cause. Second, there is the fundamental right to genomic expression. Its corollary is the fundamental wrong of social eugenics, of governmental genetic engineering, of Nazism. Third, there is the need to control happenstance conception, the unintentional creation of life. We are too valuable, too precious, to be created by force, by accident, or by chance. Unintended pregnancy is a disease to be cured, not a condition to be suffered. Finally, there is the bioethic of respect. If we do not teach respect for diversity, then we will suffer its loss. In a biotechnical age, we either block discrimination or become accomplices in its democidal debris.

The four bioethics of birth can guide us through this brave new world of test-tube pregnancies, technical parents, and transgenic progeny. Many of us may ask, can't this trip be avoided? The specter of cogenital marriages with hybrid human–animal children formed from semen masturbated twenty years earlier is—well—too much! The bioethical trip cannot be avoided, because our sociobiological environment has changed too much. Our own evolutionary processes have produced the Age of Genomics. As H. Tristram Engelhardt, Jr., of the Baylor College of Medince, has observed: "The general capacity of humans through rational contrivances to use their physiological capacities in novel ways in order to support their reproductive goals is a capacity produced through natural selection. It has served to maximize the reproductive success of humans." Dr. Engelhardt, who is also editor of the *Journal of Medicine and Philosophy* and author of the first treatise on bioethics, concludes: "Evolution is, after all, a morally blind process that has at best adapted us to environments in which we no longer live. . . . One often discovers, with chagrin, that one's most heartfelt convictions are indefensible prejudices." In other words, thriving in the Age of Genomics requires discarding the biases of our past. Judging people's worth based on their genomically directed phenotype, their body, will be primitive. Evaluating people based on how they came into this world will be obscene.

We've braved the reproductive choice of contraceptive pills

and come out a stronger, more diverse community. The female half of us is now increasingly empowered to share its multitudinous gifts in the workaday world. Tens of millions of people have turned from a life of bearing one child after another to a life of enjoying the sensations of their bodies and the activities of their minds.

We've braved the mandatory biochemistry of inoculative vaccinations, neonatal blood tests, and chlorinated water. The result has been a healthier, more diverse community. For example, many of us might have died because our biochemistry was more susceptible to the ravages of diseases against which we are all now inoculated. Similarly, the government doesn't tell us how much or when to drink water. It just ensures that the water we drink is safe. The government doesn't take our children away from us. It just lets us know that this child's blood indicates a need for special help.

Even in the area of transgenics, there is so much to welcome and very little to fear. First, we must remember that we are all somewhat transgenic to start with—about 98 percent of our genes are no different from the ones we inherited from our cousins the chimpanzees. Pig hearts work very similarly to human hearts, and horse hormones are hardly different from our own. So it is really irrational to fear the intentional creation of new blends of animal and human chromosomes—that kind of thing has been going on for millennia. Now we might not want to think of a horse that talks or a winged man, but we have been on these journeys before. The notion of going anywhere we go in our full metal horsepower, or of spending days of our lives jetting above the clouds—these thoughts were as alien to our ancestors as is the thought of transgenics to us. We've braved the frontiers of cars and planes and won a tremendously more enjoyable and productive life in the process. We can brave the frontiers of genomics and win a similar expansion of life.

Faced with new epochs, like the Age of Genomics, we have but three choices. We can close the door on the epoch as not worth the social changes we must make in accommodation. This approach throws out the good and the bad. It is about as attractive as a return to the Dark Ages. We can open the door to the epoch, but use the social system of the past to deal with the tools of the future. This approach welcomes both the good and the bad. It is exemplified best by what happened when Germany's age-old racial superiority rhetoric was combined with the new tools of industrialization. The result was mass-produced misery and assemblyline death.

Finally, we can open the door to the new epoch, but consciously change our society, too, so that we like the new world we create. This approach welcomes the good and discards the bad. It does take more effort and foresight, but it also avoids much pain and suffering. It takes open-mindedness, but it avoids capriciousness. To achieve diversity, we must give up conformity.

The changes we make to society must be the ones that enable each soul to be no less free and no less happy. This is exactly what is accomplished with the bioethics of birth. We carry our procreative freedom into the Age of Genomics intact. And where the scope of government expands, it is only to free female bodies from the disease of unwanted gestation, and to free all souls from the plague of unfair discrimination. The bioethics of birth will ensure that the Human Genome Project spawns many genies of life.

afterword coming next: an age of euthenics

When we are planning for posterity, we ought to remember that virtue is not hereditary.

THOMAS PAINE
Common Sense, 1776

Any book that focuses on the seeds of sex unavoidably fails to do justice to soil, sunlight, and nurturing. Our demographic circumstances outweigh our demographic characteristics. There is a much more meaningful field than eugenics. It is called "euthenics," the science of improving the physical and intellectual capacities of humanity by control and improvement of living conditions.

Why isn't more attention paid to euthenics than eugenics? Probably because humans are suckers for easy solutions. More people pop diet pills than do aerobics. More people watch television than create their own entertainment. Not surprisingly, people are more willing to attribute human characteristics to inbred genetic factors than to personal discipline.

Perhaps it is just natural to wish for the day when we can improve our children's abilities by ticking off characteristics on a form in the obstetrician's office. It certainly is easier than sitting down to do homework with the kids, teaching them to type, or learning sign language with them. Aggregating all our laziness on the national scale, we find budget-cutters eager to slash spending on education and urban renewal while funding the Human Genome Project's quest for a genetic Holy Grail. Genies are so attractive because they are so easy. Just rub the bottle and get what you want. How much more difficult to build that magic palace from scratch!

One of the fundamental fallacies of eugenics is: Making "better people" will make a better world. I've been to lots of ivory towers. They are nice places to visit, but I wouldn't want to live in one. The message of euthenics makes much more sense: Making a better world makes a better world. Genes don't offer a sense of accomplishment. Only accomplishment itself offers this bedrock *raison d'être*. And the *magnitude* of accomplishment depends not at all on one's genetic starting point. Whether we climb from a canyon to a cliff or from a meadow to a mountain peak, it is the climb, celebrated by society, that makes our world a wondrous place.

The moral of every genie-in-the-bottle story is "Beware of what you wish." And so it is with eugenics. It is doubtful that more geniuses or Olympians will bring us relief from warfare, poverty, or disease. To the contrary, our armies, economies, and public health systems are already (mis)managed by men and women of high intelligence. It is likely, however, that our wishing for better genetics will lead to a "genism" every bit as pernicious as sexism, racism, and nationalism. It is also likely that we will be deflected from a eudaemonic ethos of rationally building a better world while we are concentrating our attention on wistfully creating "better people." In short, if we wish for a class of smarter or stronger people, we will probably get them. But we will probably be sorry for the consequences of that wish—a powerful microclass of Olympians who pay scant attention to improving the lives of billions.

We cannot wish for euthenics because genies don't deliver processes, they deliver things. Euthenics is something that we get to create with our own hands and determination. In an Age of Euthenics, the prevailing ethic will be one of improving the world around us, and of leaving no one too far behind. The focus will be not on "what did I get," but on "what did I do." In an Age of Euthenics, genies are largely irrelevant. What you started with is interesting only to measure how far you have gone.

Our policy challenge is not the biological smorgasbord of genomics, but the biological reductionism of society. We already enjoy a wide variety of genomic diversity, and the ability to augment and build upon that diversity is a wonderful new capability. Today's society, however, makes a serious error by attributing any sort of personal attributes to biological features. When we stereotype people who look a certain way, outwardly or genomically, as having certain personal characteristics we engage in a dangerously lazy mode of thinking. Whether we reduce people to their skin tone—or to their genome—we are wrong.

We must now understand that the car is not the vacation, and the body is not the soul. We must mature a lot as a society in order to enter the Age of Genomics without demographic disaster. And we must mature now, because the demographics of death lie at the doorstep and the holocaust of sex is already in the hall. I believe this evolution in human mindset is achievable, even in a very short span of years.

During the second half of the twentieth century, a great deal of progress has been achieved in creating a multicultural world. We are, of course, still a long way from its realization. Nevertheless, the ability to deal with all sorts of people disembodied from their biology, as in cyberspace, has facilitated the debunking of stereotypic modes of thinking. The globalization of mass media has also helped to debunk biological stereotypes—even as it promotes them—because similar-looking people are shown to behave in so many different ways that no one stereotype can achieve much credence. Finally, the ever more insistent enforcement of nondiscrimination legislation enables greater numbers of diverse people to enter into personal contact in workplace, educational, and recreational surroundings. Increasingly we have the opportunity to see that a person's degree of melanin, physical ability, or sexual conformity is quite irrelevant to her or his personality. We are ready to understand that a person's genome will not dictate personality. That's a tremendous accomplishment for a society still struggling out of a several-thousand-year history of continuous ethnic warfare.

Charles Darwin observed that "the highest possible stage in moral culture is when we recognize that we ought to control our thoughts." It is still automatic for us to see a body and think a stereotype. But we are also increasingly able to control that thought and to act independently of it. We are increasingly able to separate the person from the body, the soul from the genome. Society as a whole has not yet achieved Darwin's "highest possible stage," but its terrain is in sight. We may hope that we are close enough so that the Age of Genomics, operating under our new bioethics of birth, will yield biotechnical building blocks for health and happiness rather than stereotypical fodder for social control.

The biotechnology revolution should be beneficial so long

as we control our own thoughts while we manipulate our own genomes. This calls for constant education, accurate information, and a free flow of communication. In essence, euthenistic improvements in our society's infrastructure—adequate schools, relief from poverty, strengthening of community—are exactly what is needed in order to reap the benefits of genomics without suffering its misuse. So let a call go out from this day forward that genomics without euthenics is a fraud. Our fate is neither in the stars nor in our cells, but in our collective will to make living a wonderful experience for all people, whether alive or as yet just seeds.

appendix
and
index

appendix
international documents on genomic rights

This Appendix includes three different approaches to encompassing the bioethics of birth within a binding legal instrument. The first document, "International Bill of Genomic Rights," is a statement I wrote over the past few years to convert the more academic "bioethics of birth" into a manifestolike statement of specific social rules. Its purpose is to guide governments, ethicists, and activists in developing new legislation.

From 1993 through 1996 many genomic law proposals were presented to the Bioethics Subcommittee of the Law and Medicine Committee of the International Bar Association (IBA). As might be expected of a committee process, especially an international committee of lawyers, the various proposals were transmuted into the more sedate legal principles found in the second document, "Draft International Convention on the Human Genome." This document was approved by the Bioethics Subcommittee of the IBA at its fourth session in Berlin, Germany, during October 1996. Its purpose is to serve as a spur for diplomats to develop common international legal standards for genomic activity worldwide.

Finally, the third document, "Preliminary Declaration on the Human Genome and Human Rights," is the result of a consensus process involving dozens of bioethicists, jurists, and scientists from around the world meeting within the International Bioethics Committee of the United Nations Educational, Scientific and Cultural Organization (UNESCO). Its purpose is to position the human genome within the framework of human rights law and to ensure common bioethical standards worldwide. It will be proposed for international approval during the late 1990s.

Each of the three genomic documents has its own benefits and drawbacks. What those relative differences are, however, is very much in the eye of the beholder. What do they have in common? The four bioethics of birth explained in this book: the human genome as the common heritage of us all, the freedom of personal genomic choice, the prohibition of forced genomic actions, and the obligation to eliminate genomic discrimination.

INTERNATIONAL BILL OF GENOMIC RIGHTS

HAVING CONSIDERED the provisions of Article 12 of the 1966 International Covenant on Economic, Social, and Cultural Rights (*entered into force in 1976*), which recognizes the right of everyone to the enjoyment of the highest attainable standard of physical and mental health and that the necessary steps to achieve this include the prevention of diseases and the creation of conditions that assure to all medical service in the event of sickness;

RECOGNIZING that understanding of the human genome and the development of human genome technology will materially assist in the prevention or cure of congenital and environmentally acquired diseases and will help maximize the availability of medical services through advances in predictive medicine;

FURTHER CONSIDERING the provisions of Articles II and III of the 1948 Convention on the Prevention and Punishment of the Crime of Genocide (entered into force, *January 12*, 1951), which proscribe the imposition of measures intended to prevent births within demographic groups;

RECOGNIZING that human genome information can be used to prevent births within certain demographic groups, to develop biological weapons that can selectively injure or kill members of certain demographic groups, or to alter inheritable human characteristics in a manner unrelated to medical necessity;

REMEMBERING that at all periods of history claims of genetic superiority have created great conflict for society and inflicted great losses on humanity;

Being convinced that, in order to ensure humanity benefits from the medical and health advances made possible by a fuller understanding of the human genome, without falling prey to odious or malicious applications of the technology, it is necessary to have international cooperation in the regulation of uses of human genome information;

It is hereby declared:

Article I

The human genome in the aggregate, and each unique human genome in particular, are the common heritage of humanity, and are correspondingly deserving of special legal recognition.

Human genomes contain an invaluable biochemical record of our ancient evolution and contain the blueprint for countless future generations. Legal recognition of the human genome is necessary as a matter of intergenerational equity.

Article II

Governments shall make no law abridging the freedom of individuals to create human genomes for the intentional formation of family life.

The "intentional formation of family life" is the act of creating, or directing the creation of, a new genome by combining two different contributed human half-genomes, modified or unmodified by the substitution of other human or nonhuman gene sequences, with the objective of producing a new and unique member of a small social group united by mutual love and legal responsibilities. Human genomes cannot be created anonymously or by corporate entities, but neither can numerical or demographic limits be imposed on the parents of a genome.

Article III

Human genomes shall have the right, but not the obligation, to benefit from the use of new biotechnological techniques, including the use of modified genetic sequences, as directed by the bearers of such genomes.

No persons are obligated to perform genetic screening or genetic surgery upon a new genome they produce, but neither may any persons be prevented from performing genetic screening or genetic surgery. Governments are obligated to ensure all their citizens have the ability to exercise their genomic rights regardless of their economic resources.

Article IV

No human genome shall be forced into life with any particular genomic characteristics other than those selected for it in the free exercise of parental judgement.

Because human genomes should only be created by individual human beings in an act of love, no one may be coerced into producing genomes with any particular characteristics.

Article V

No human being shall be denied any civil, criminal, or health care rights on the basis of its genomic characteristics.

Because the human genome is the common heritage of all humanity, no discrimination should be inflicted upon people based upon their share of humanity's heritage.

Article VI

Governments shall take all necessary steps to ensure, to the greatest extent feasible, that any human genome that occurs is intentionally formed.

Governments may implement universal inocuseeding as a means of ensuring the creation of intentional life.

DRAFT INTERNATIONAL CONVENTION ON THE HUMAN GENOME

The States Parties of this Convention,

Inspired by the prospects of great improvements in health care as a result of technological advances in understanding the human genome,

Recognizing the common interest of all humankind in the progress of the development of human genome technology,

Believing that the decoding of the human genome, and the development of human genome technology, should be carried on for the benefit of all peoples irrespective of the degree of their economic or scientific development,

Desiring to contribute to broad international cooperation in the scientific as well as the ethical aspects of the use of human genome technology,

Believing that such cooperation will contribute to the development of mutual understanding and to the strengthening of friendly relations between States and peoples,

Believing that genomic research and genetic medicine can lead to significant alleviation of human suffering and to improvement of health care and the quality of life for humankind,

Encouraging continued support for the advancement of genomic research and its applications, subject to appropriate safeguards,

Bearing in mind that prevention of births on the basis of genomic information with the intention of wholly or partially eliminating national, ethnical, racial or religious groups would be an international crime under the Convention on the Prevention and Punishment of the Crime of Genocide of December 9, 1948,

Bearing in mind that discrimination on the basis of genomic information would be inconsistent with the principles established in the International Convention on the Elimination of all Forms of Racial Discrimination of March 7, 1966, which condemns any distinction, exclusion, restriction or preference based on race, color, descent, or national or ethnic origin,

Convinced that a Convention on the Human Genome will further the goal of providing all people with the highest attainable standard of health care while minimizing the risks of new forms of discrimination based on genomics,

Have agreed on the following:

Article I: The Human Genome

1. The human genome is part of the common heritage of humankind.

2. Human genome technology shall be developed and used only in full and complete consistency with the common interests of humanity.

Article II: Germline Therapy and Eugenics

1. States Parties agree not to permit the performance of any act which directly or indirectly facilitates any eugenic practices using human genome technology contrary to the principles established in the International Convention on the Elimination of all Forms of Racial Discrimination of March 7, 1966.

2. Germline therapy in humans shall be proscribed except where there is indisputable proof, in accordance with international scientific standards, of the benefits and safety of such therapy.

Article III: Genomic Information and Pregnancy

Each State Party shall take effective measures, including legislation where appropriate,

(a) To ensure that pressure based on genomic rationales is not brought upon any parent of an unborn child for a pregnancy to be terminated,

(b) To proscribe, as a ground for encouraging termination of pregnancy, alleged congenital or otherwise gene related deficiencies based on any preconceived "standard of normality,"

(c) To proscribe, as a ground for encouraging termination of pregnancy, genetic predisposition to illness which may be likely to occur in later life,

(d) To impose no obligation upon persons within its jurisdiction to undergo genetic operations and to impose no financial or social consequences on persons who refuse to undergo genetic operations.

Article IV: Genetic Testing

1. Each State Party shall ensure, through legislation or other means as appropriate, that individuals, laboratories or institutions of any kind involved in genetic testing are licensed or registered by the appropriate governmental authority within whose jurisdiction they operate in accordance with internationally accepted standards of scientific accuracy, confidentiality of information, and bioethics.

2. Each State Party undertakes to guarantee, through legislation or other means as appropriate, that when genetic testing of a particular person within their jurisdiction reveals that such person has a predisposition to suffer disease or disability

in the future, then such person shall have the right exercised by freedom of choice whether to be informed of the results of such testing.

Article V: Genetic Privacy and Discrimination

1. Each State Party shall encourage the complete confidentiality of human genome information consistent with the highest internationally accepted standards of medical confidentiality and bioethics.

2. Each State Party shall prohibit and bring to an end, by all appropriate means, including legislation as required by circumstances, discrimination on the basis of genomic characteristics by any persons, group or organization where such discrimination has the purpose or effect of nullifying or impairing the recognition, enjoyment or exercise, on an equal footing, of human rights and fundamental freedoms in the political, economic, social, cultural or any other field of public life.

3. States Parties undertake:

(a) to adopt immediate and effective measures particularly in the fields of teaching, education, culture and information, with a view to combating prejudices based on genetic characteristics;

(b) to promote the scientific fact that the human genome is expressed differently according to the environment, education, living conditions and state of health of each family and each individual; and

(c) to support sensitivity to the needs of, and solidarity with all persons whose genomic inheritance creates special difficulties.

Article VI: Intellectual Property

1. The human genome in its natural state is not subject to private, national or transnational ownership by claim of right, patent or otherwise.

2. Intellectual property based upon the human genome may be patented or otherwise recognized in accordance with national laws and international treaties.

3. The collection, distribution and use of human genomic materials and associated information shall be undertaken on a basis which reflects an interest of the original source and of the depositor of the material to an equitable share of the economic benefit of commercialization based upon:

(a) use of the material and associated information;

(b) the relative significance and/or unique nature and/or rarity of the genomic characteristics of the material and associated information; and

(c) the original source and the depositor's relative contributions to the overall creation and commercial development of relevant intellectual property.

Article VII: Genomic Research

1. Each State Party shall ensure, through legislation or other means as appropriate, that all human genome research carried on within its jurisdiction is conducted in accordance with internationally accepted medical, scientific and bioethical standards.

2. Each State Party shall encourage:

(a) the availability and exchange, within the international scientific community, and in particular with developing countries, of human genomic material and associated information for the purposes of research and to develop new information about the human genome, subject to appropriate international safeguards and protection against abuse; and

(b) the availability and exchange of human genomic material and associated information through depositories meeting internationally accepted standards, with the objective that the collection, authentication, storage and distribution of the materials and associated information for research will be conducted in a manner which is consistent with this Convention and with such internationally accepted standards.

Article VIII: International Cooperation

States Parties shall undertake, with appropriate protection of intellectual property rights, to foster the international dissemination of scientific knowledge concerning the human genome and to foster scientific and cultural cooperation, particularly between industrialized and developing countries.

Article IX: Interpretation

No provision of this Convention may be used by any State, group or person to ends contrary to the principles set forth herein.

Article X: Compliance

1. Each State Party shall annually report and publish its progress in complying with the terms of this Convention.

2. Each State Party agrees to support efforts on behalf of the national law associations, in consultation with scientific, medical and other relevant organizations, independently to collect information concerning worldwide compliance with the provisions of this Convention.

3. Each State Party agrees to support its national law asso-

ciations to participate in meetings of the International Bar Association dedicated to assessing worldwide compliance with the provisions of this Convention.

4. Each State Party agrees to provide due consideration to biennial reports and recommendations of the International Bar Association with regard to worldwide compliance with this Convention.

Article XI: Signature, Ratification and Accession

1. This Convention shall be open to all States for signature. Any State which does not sign this Convention before its entry into force in accordance with paragraph 3 of this article may accede to it at any time.

2. This Convention shall be subject to ratification by signatory States. Instruments of ratification and instruments of accession shall be deposited with the Governments of the United States of America, the United Kingdom of Great Britain and Northern Ireland and Japan, which are hereby designated the Depository Governments in deference to their respective jurisdiction over the three global headquarters of the Human Genome Organization.

3. This Treaty shall enter into force upon the deposit of instruments of ratification by five Governments including the Governments designated as Depositary Governments under this Convention.

4. For States whose instruments of ratification or accession are deposited subsequent to the entry into force of this Convention, it shall enter into force on the date of the deposit of their instruments of ratification or accession.

5. The Depositary Governments shall promptly inform all signatory and acceding States of the date of each signature,

the date of deposit of each instrument of ratification of and accession to this Convention, the date of its entry into force and other notices.

6. This Convention shall be registered by the Depositary Governments pursuant to Article 102 of the Charter of the United Nations.

Article XII: Amendments

Any State Party to this Convention may propose amendments to this Convention. Amendments shall enter into force for each State Party to the Convention accepting the amendments upon their acceptance by a majority of the States Parties to the Convention and thereafter for each remaining State Party to the Convention on the date of acceptance by it.

Article XIII: Periodic Review

Ten years after the entry into force of this Convention, the question of the review of this Convention shall be included in the provisional agenda of the United Nations General Assembly in order to consider, in the light of past application of the Convention, whether it requires revision. The International Bar Association and other relevant international organizations are invited to produce reports and recommendations on the subject of any necessary revisions. However, at any time after the Convention has been in force for five years, and at the request of one third of the States Parties to the Convention, and with the concurrence of the majority of the States Parties, a conference of the States Parties shall be convened to review this Convention. The International Bar Association, and other relevant international organizations, shall be invited to attend this conference in the role of an expert advisor to the State Parties.

Article XIV: Withdrawal

Any State Party to this Convention may give notice of its withdrawal from the Convention one year after its entry into force by written notification to the Depository Governments. Such withdrawal shall take effect one year from the date of receipt of this notification.

Article XV: Languages

This Convention, of which the English, Russian, French, Spanish and Chinese texts are equally authentic, shall be deposited in the archives of the Depositary Governments. Duly certified copies of this Convention shall be transmitted by the Depositary Governments to the Governments of the signatory and acceding States.

PRELIMINARY DRAFT OF A UNIVERSAL DECLARATION ON THE HUMAN GENOME AND HUMAN RIGHTS

The General Conference,

Recalling that the Preamble of UNESCO's Constitution refers to "the democratic principles of the dignity, equality and mutual respect of men," rejects "the doctrine of the inequality of men and races," stipulates "that the wide diffusion of culture, and the education of humanity for justice and liberty and peace are indispensable to the dignity of men and constitute a sacred duty which all the nations must fulfil in a spirit of mutual assistance and concern," proclaims that "peace must be founded upon the intellectual and moral solidarity of mankind," and states that the Organization seeks to advance "through the educational and scientific and cultural relations of the peoples of the world, the objectives of international peace and of the common welfare of mankind for which the United Nations Organization was established and which its Charter proclaims,"

Solemnly recalling its attachment to the universal principles of human rights, affirmed in particular in the Universal Declaration of Human Rights of 10 December 1948 and in the two International United Nations Covenants on Economic, Social and Cultural Rights and on Civil and Political Rights of 16 December 1966, in the United Nations Convention on the Prevention and Punishment of the Crime of Genocide of 9 December 1948, the International United Nations Convention on the Elimination of All Forms of Racial Discrimination of 21 December 1965, the United Nations Convention on the Elimination of All Forms of Discrimination Against

Women of 18 December 1979, the United Nations Convention on the Rights of the Child of 20 November 1989, the Convention on the Prohibition of the Development, Production and Stockpiling of Bacteriological (Biological) and Toxin Weapons and on their Destruction of 16 December 1971, the UNESCO Convention against Discrimination in Education of 14 December 1960, the UNESCO Declaration of the Principles of International Cultural Co-operation of 4 November 1966, the UNESCO Recommendation on the Status of Scientific Researchers of 20 November 1974, the UNESCO Declaration of Race and Racial Prejudice of 27 November 1978 and the ILO Convention (N° III) concerning Discrimination in Respect of Employment and Occupation of 25 June 1958,

Bearing in mind the international instruments which could have a bearing on the applications of genetics in the field of industrial property, *inter alia,* the Bern Convention for the Protection of Literary and Artistic Works of 9 September 1886 and the UNESCO Universal Copyright Convention of 6 September 1952, as last revised in Paris on 24 July 1971, the Paris Convention for the Protection of Industrial Property of 20 March 1883, as last revised at Stockholm on 14 July 1967, and the Budapest Treaty of the WIPO on International Recognition of the Deposit of Micro-Organisms for the Purposes of Patent Procedures of 28 April 1977,

Bearing in mind also the United Nations Convention on Biological Diversity of 2 June 1992 and *emphasizing* in that connection that the recognition of the biological diversity of humanity should not give rise to any interpretation of a social or political nature which could call into question the fundamental principle of equal dignity inherent in all members of the human family,

Recalling 22 C/Resolution 13.1, 23 C/Resolutuon 13.1, 24 C/Resolution 13.1, 25 C/Resolutions 5.2 and 7.3, 27 C/Resolution 5.15 and 28 C/Resolutions 0.12, 2.1 and 2.2, urging UNESCO to promote and develop ethical studies, and the actions arising out of them, on the consequences of scientific and technological progress in the fields of biology and genetics, within the framework of respect for human rights and freedoms,

Recognizing that:

(a) Research on the human genome and the resulting applications open up vast prospects for progress in improving the health and well-being of individuals and of humankind as a whole,

(b) The applications of genetic research must, however, be regulated in order to guard against any eugenic practice that runs counter to human dignity and human rights,

(c) The results of research on the human genome should in no case be used towards military or bellicose ends,

(d) The human and social situations generated by advances in biology and genetics require that there should be a very open international debate, ensuring the free expression of the various shades of socio-cultural, religious and philosophical opinion,

Considering that the principles relating to the human genome and the protection of the individual based, in accordance with the preamble to the Universal Declaration of Human Rights, on "recognition of the inherent dignity and of the equal and inalienable rights of all members of the human family (which) is the foundation of freedom, justice and peace in the world,"

Proclaims that the human genome is the common heritage of humanity and hereby *adopts* the principles set forth in the present Declaration.

A. THE HUMAN GENOME

Article 1

The human genome is a fundamental component of the common heritage of humanity.

Article 2

(a) The genome of each individual represents his or her specific genetic identity.

(b) An individual's personality cannot be reduced to his or her genetic characteristics alone.

(c) Everyone has a right to the respect of their dignity and of their rights regardless of these characteristics.

Article 3

The human genome, which is by nature evolutive and subject to mutations, contains potentialities that are expressed differently according to the environment, education, living conditions and state of health of each family and each individual.

B. RESEARCH ON THE HUMAN GENOME

Article 4

The protection of the individual with respect to the implications of research in biology and genetics is designed to safeguard the integrity of the human species, as a value in its own right, as well as the respect for the dignity, freedom and the rights of each of its members.

Article 5

(a) Research, which is an essential activity of the mind, has the function, in the fields of biology and genetics, of advancing knowledge, relieving suffering and improving the health and well-being of the individual and of humankind as a whole.

(b) Everyone has the right to benefit from advances in biology and genetics, with due regard to his or her dignity and rights.

Article 6

No scientific advances in the fields of biology and genetics should ever prevail over the respect for human dignity and human rights.

C. INTERVENTIONS AFFECTING THE HUMAN GENOME

Article 7

No intervention affecting an individual's genome may be undertaken, whether for scientific, therapeutic or diagnostic purposes, without rigorous and prior assessment of the risks and benefits pertaining thereto and without prior, free and informed consent of the person concerned or, where appropriate, of his or her duly authorized representatives, guided by the person's best interests.

Article 8

No one may be subjected to discrimination on the basis of genetic characteristics that aims or has the effect of injuring the recognition of human dignity or the enjoyment of his or her rights on the grounds of equality.

Article 9

The confidentiality of genetic data associated with a named person and stored or processed for the purposes of research or any other person, must be protected from third parties.

Article 10

Every individual has the right to just reparation for any injuries sustained as a direct result of an intervention affecting his or her genome.

D. RIGHTS AND OBLIGATIONS OF RESEARCHERS

Article 11

States shall ensure the intellectual and the material conditions favourable to research on the human genome, in so far as this research contributes to the advance of knowledge, the relief of suffering and the improvement of the health and well-being of the individual and of humankind as a whole.

Article 12

States shall provide a framework for research with due regard for democratic principles, in order to safeguard the dignity and rights of the individual, to protect public health and the environment.

Article 13

In view of its ethical and social implications, research in biology and genetics entails special responsibilities as regards the meticulousness, caution and intellectual honesty required of researchers.

E. DUTIES AND RESPONSIBILITIES TOWARDS OTHERS

Article 14

States must guarantee the effectiveness of the duty of solidarity towards individuals, families and population groups that are particularly vulnerable to disease or disability linked to anomalies of a genetic character.

Article 15

States shall recognize the value of promoting, at various appropriate levels, the establishment of independent, multidisciplinary and pluralist ethics committees to identify ethical, social and human issues raised by research and interventions affecting the human genome.

F. INTERNATIONAL CO-OPERATION

Article 16

States shall undertake, with due regard for democratic principles, to foster the international dissemination of scientific knowledge concerning the human genome and to foster scientific and cultural cooperation, particularly between industrialized and developing countries.

Article 17

States shall undertake to promote specific teaching and research concerning the ethical, social and human foundations and implications of biology and genetics.

Article 18

States shall undertake to encourage any other form of research, training and information conducive to raising the awareness of society of its responsibilities regarding the basic choices entailed by advances in biology and genetics.

G. IMPLEMENTATION OF THE DECLARATION

Article 19

States shall undertake to ensure that the principles set out in this Declaration are respected.

Article 20

The principles set out in this Declaration shall guide all authorities and other persons responsible for their implementation.

Article 21

States shall undertake to promote, through education, training and information, respect for the aforementioned principles, based on human dignity and human rights and to foster their recognition and effective application.

Article 22

The International Bioethics Committee of UNESCO shall monitor observance of the principles set out in this Declaration. For this purpose, it may make recommendations and give advice.

Article 23

No provision of this Declaration may be used by any State, group or person to ends contrary to the principles set forth herein.

index

SUNY BROCKPORT
3 2815 00809 0162